The Divine Proportion

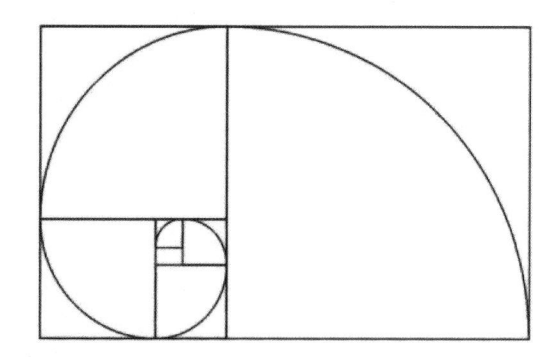

SCOTT ONSTOTT

Copyright © 2015 SIPS Productions Inc.
All Rights Reserved.

ISBN 978-1516964390

Learn more at http://www.secretsinplainsight.com

Introduction

For many years mathematicians, physicists, astronomers, philosophers, biologists, architects, historians, musicians, psychologists, artists and mystics have been interested in the divine proportion—and by the fact that you have picked up this book—it would seem you share their interest.

The divine proportion has fascinated thinkers throughout the ages because of its unique aesthetic, mathematical and scientific properties. Originally called the "extreme and mean ratio" by Euclid, it has also been called the divine section, golden ratio, golden mean, golden number and many other names. All of these refer to the act of cutting a line at one uniquely important point.

Point C cuts the line AB at the divine proportion if and only if the whole line ($\alpha+\beta$) is in proportion to the longer segment (α) in exactly the same proportion as the longer segment (α) is to the shorter segment (β).

$$\frac{\alpha+\beta}{\alpha} = \frac{\alpha}{\beta} = \Phi \approx 1.618033989$$

The singular act of cutting the line at point C sets up a self-similar ratio that is fractally proportioned the same above as it is below. This process can be repeated and extended to triangles, squares, pentagrams and many other shapes. Amazingly, the implications of the divine proportion go far beyond geometry.

The divine proportion has a non-repeating decimal expansion that is traditionally represented by the Greek letter Φ, and "phi" has mathematical properties that will continue forever to mystify:

$$\Phi = \sqrt{1+\sqrt{1+\sqrt{1+\sqrt{1+\sqrt{1+\sqrt{1+\ldots}}}}}}$$

$$\Phi = 1 + \cfrac{1}{1+\cfrac{1}{1+\cfrac{1}{1+\cfrac{1}{1+\cfrac{1}{1+\cfrac{1}{1+\cfrac{1}{1+\ldots}}}}}}}$$

$$F(n) = \frac{\Phi^n - (1-\Phi)^n}{\sqrt{5}} \quad \text{Fibonacci numbers}$$

$$\Phi = \lim_{n \to \infty} \frac{F(n+1)}{F(n)} \quad \text{Ratio of successive terms of any Fibonacci-like sequence}$$

There are many geometric ways to derive Φ. The following is among the most beautiful of methods that can be constructed using a compass and straightedge. Following the construction sequence will give you a taste of sacred geometry.

In this method you won't ever change the spread of the compass, so all circles you draw will be the same size. Begin by drawing a circle. Place the point of the compass at some arbitrary point on the circumference of the existing circle, perhaps near the top, and draw another circle. The points of intersection are labeled A and B. This shape has been known for centuries as the *vesica piscis* (Latin for "fish bladder") because the outline of the almond-shaped space in the middle also resembles a fish's swim bladder. Many will more readily perceive the center of the vesica piscis as the shape of a human eye.

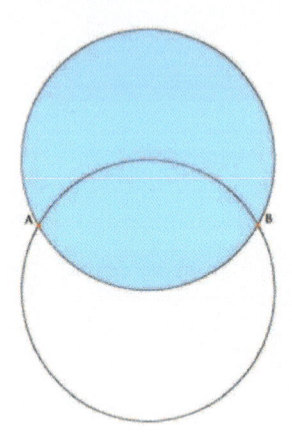

Draw two more circles using A and B as centers. The points where the new circles intersect the original circle are labeled C and D below.

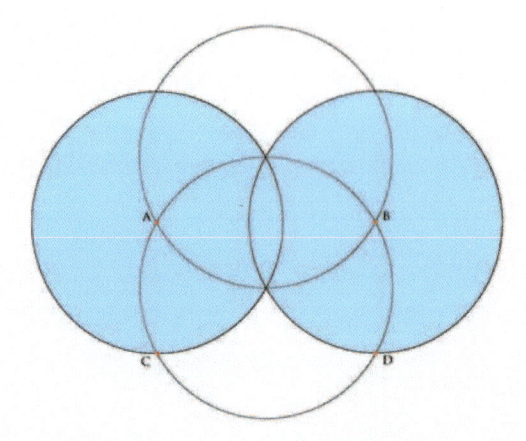

Draw two more circles using C and D as centers. One of the points where the new circles intersect is labeled E below. Draw another circle centered at E. This construction of six circles around one is traditionally called the *Seed of Life*. If you were to continue in this way, drawing new circles at earlier circle intersection points, you would end up drawing the famous *Flower of Life* diagram, but drawing the *Seed of Life* is sufficient for this exercise of creating the divine proportion.

The point where the top two circles intersect is labeled F. Use a straightedge to connect the dots and draw equilateral triangle CDF.

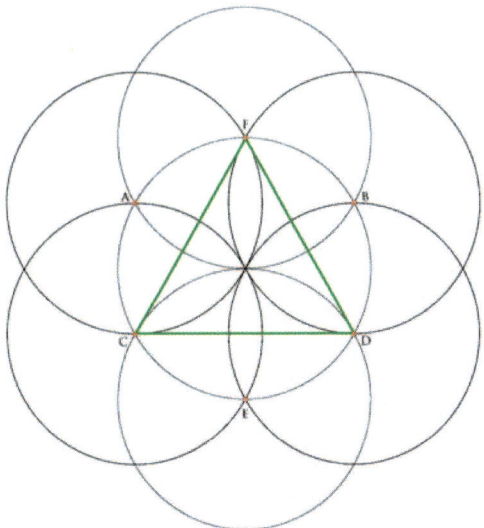

Draw lines GA and GB as shown below. These segments are automatically divided in half by the triangle at H and I. Draw a line through H and I and extend it to point J where the line intersects the original circle. The length of segment HI is labeled α, and the length of segment IJ is labeled β. Line HJ and its segments are divinely proportioned as HJ / HI = HI / IJ.

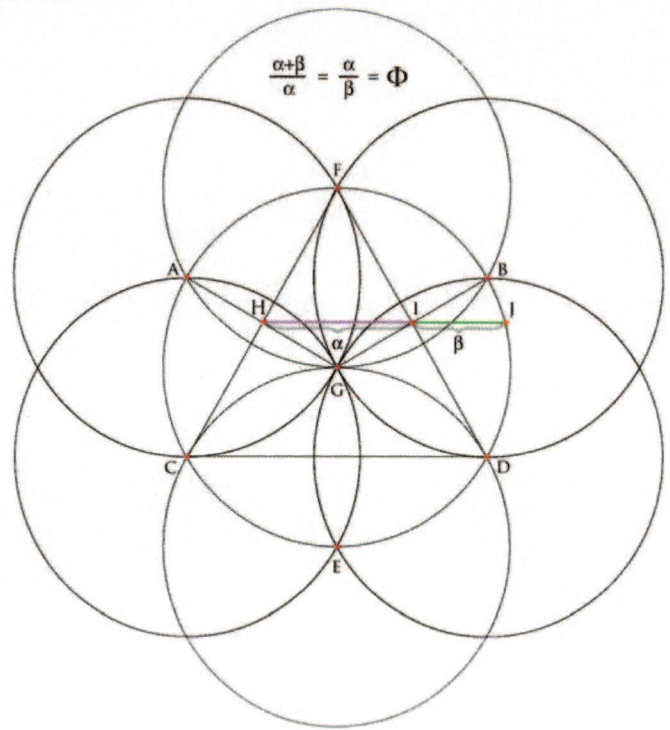

$$\frac{\alpha+\beta}{\alpha} = \frac{\alpha}{\beta} = \Phi$$

The divine proportion is found everywhere from the smallest possible measurable distance where quantum effects predominate ($\Phi \times 10^{-33}$ cm)[1], to resonating atoms[2], to the code of life in DNA, to the human form and the evolution of consciousness. The divine proportion is masterfully encoded in the Great Pyramid's location and form, in the *Eye of God* geometry, in Renaissance art and Modern architecture, in the geometry of plants and rainbows, in the harmonious relationships of a circle to square and of a cube to a sphere, in the Earth and Moon system, in planetary relationships, in time, in the structure of our galaxy and far beyond. Astoundingly, humans are divinely proportioned in the grand scale of all things!

I have been guided on this epic journey of rediscovery, and the illustrations that follow are signposts I created along my path. I hope that my illustrations will help guide you *to remember what you already know,* for the divine proportion is in your heart as it is in the heart of the universe. The book that you hold in your hands puts you in a privileged position to perceive the patterns and to feel your interconnection with all that is, was and shall be. We live in a divinely proportioned, magical universe.

1. $\Phi \times 10^{-33}$ cm measures the Planck length with 99.9% accuracy.
2. See http://bit.ly/1UQ3wIA for more information on the divine proportion at the atomic scale.

How To Use This Book

I recommend leafing through the illustrations until one in particular catches your eye. Contemplate that image for a while, appreciating its qualities and ponder what might have drawn you to the image. Turn then to the endnotes and focus on digesting the associated written information. You might then reexamine the image and see some things you missed before. Repeat this process for another image as you continue through the book.

By alternating your states of awareness between contemplation, concentration and meditation I believe you will resonate more effectively with the divine proportion. I hope that you will be as inspired as I am with this process and turn your loved ones onto this amazing path of rediscovery.

Scott Onstott
September 2015
British Columbia

List of Illustrations

1. Golden and Fibonacci Rectangles
2. Golden Pentagram
3. Phive
4. Jacob's Ladder
5. Fibonacci Sequence (mod 9)
6. Masterpieces
7. Ph-eyes
8. Golden Architecture
9. Golden Angle
10. Golden Triangle
11. Golden Violin
12. Cubing the Sphere
13. The Gap
14. Phi of God
15. Squaring the Circle
16. Sevens
17. Core
18. Epiphany
19. Golden Template
20. Birthplace of Isis
21. 3:4:5 Moon Earth
22. Body Temple
23. Location, Location, Location
24. The Nile
25. Human Development
26. Golden Timelines
27. Golden Scale
28. Spiral Galaxy
29. Saturn Earth Proportions
30. Earth Moon Proportions
31. Basis of the Metric System
32. Golden Human
33. Golden DNA

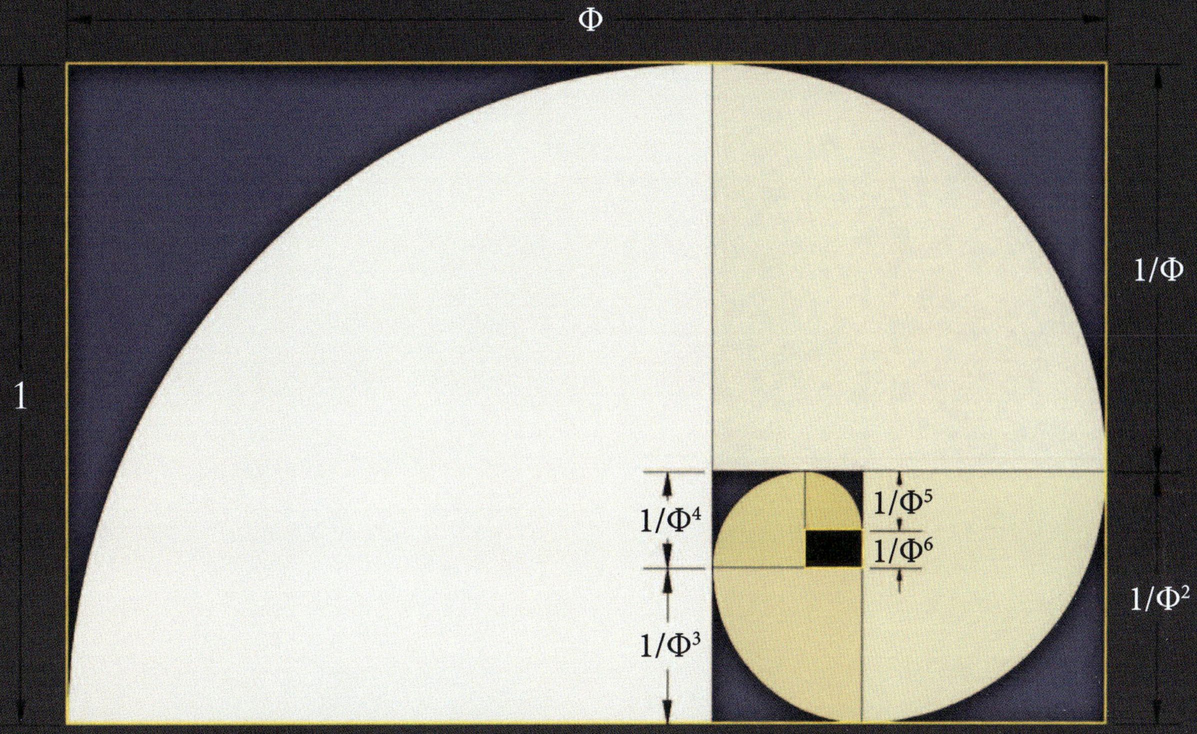
Φ
1
1/Φ
1/Φ⁴
1/Φ⁵
1/Φ⁶
1/Φ³
1/Φ²

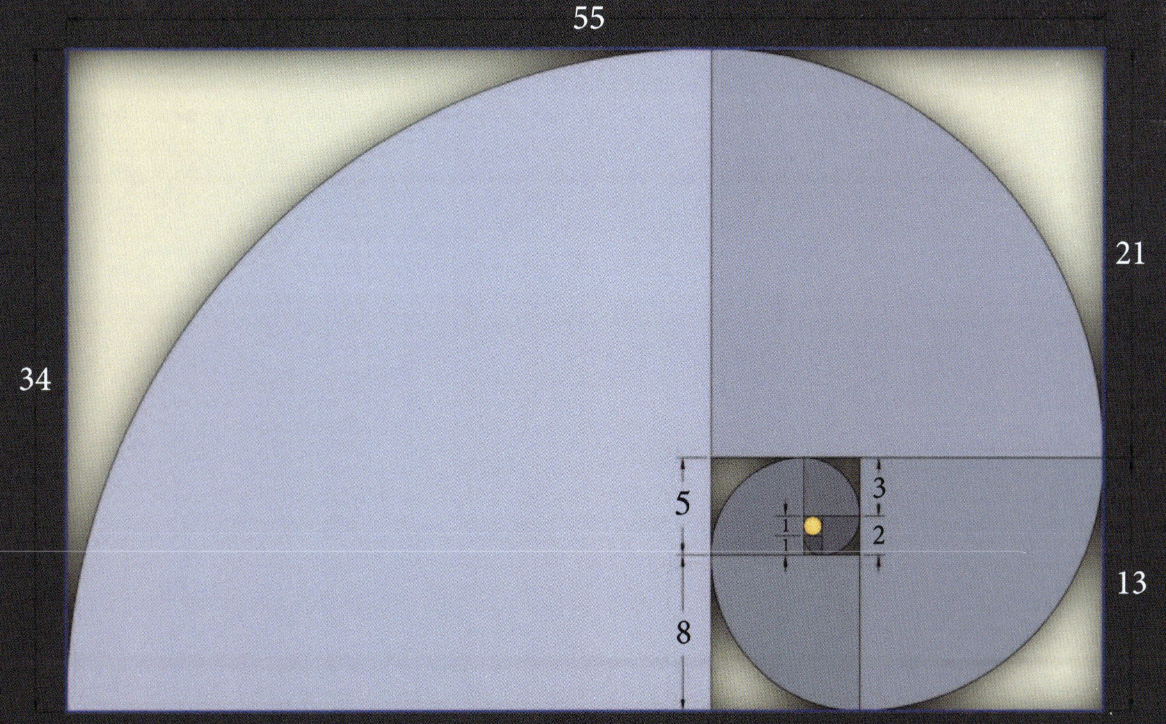
55
34
21
5
3
1
1
2
8
13

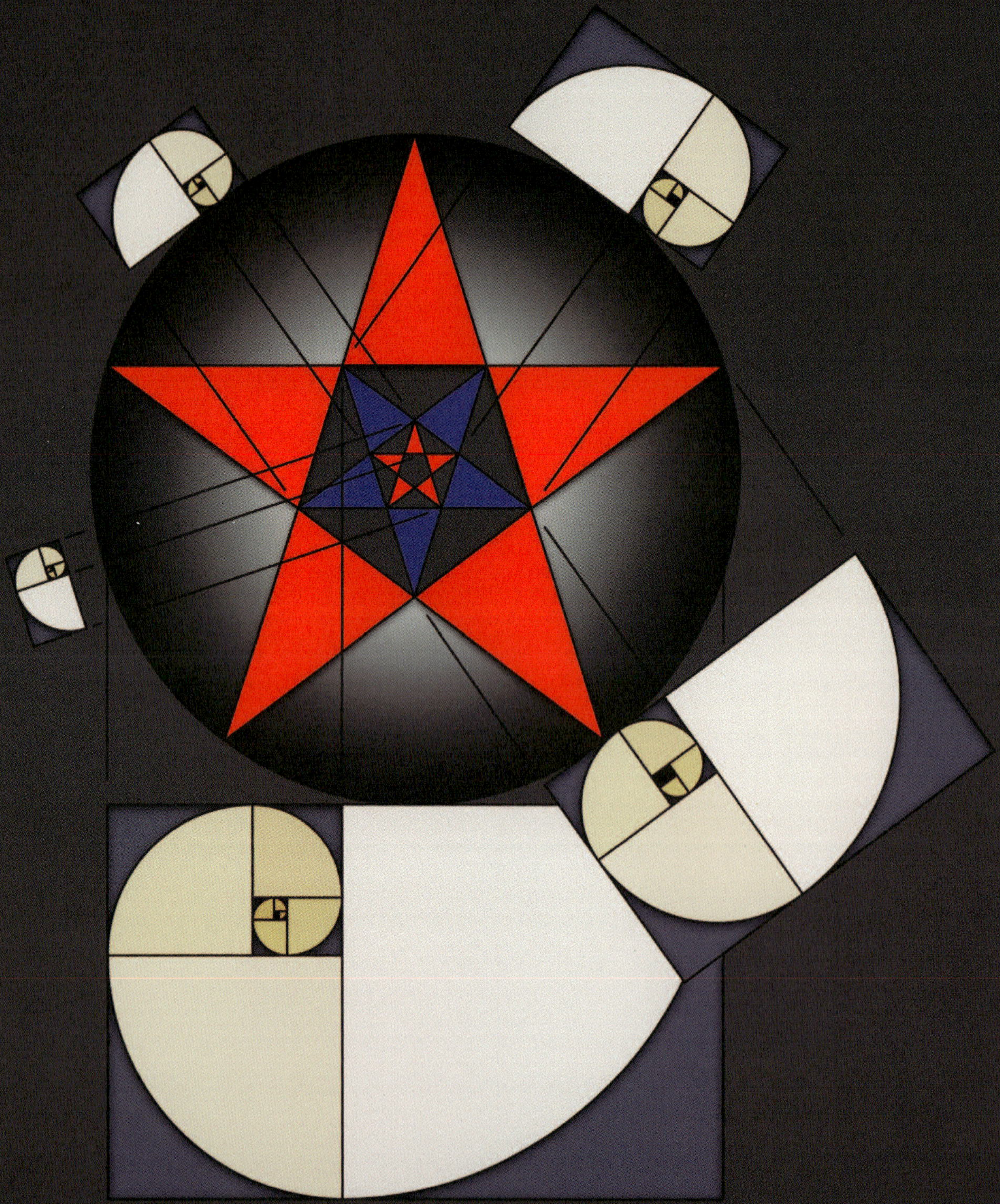

$$.5 \times 5^{.5} - .5$$

$$.5$$

$$1/\Phi = .5 \times 5^{.5} - .5 \approx 0.618033989$$
$$\Phi = .5 \times 5^{.5} + .5 \approx 1.618033989$$

89		1/Φ
55 34		
21		1/Φ² Φ
		1/Φ³
13 8		1/Φ⁴
5 3		1/Φ⁵
...		...

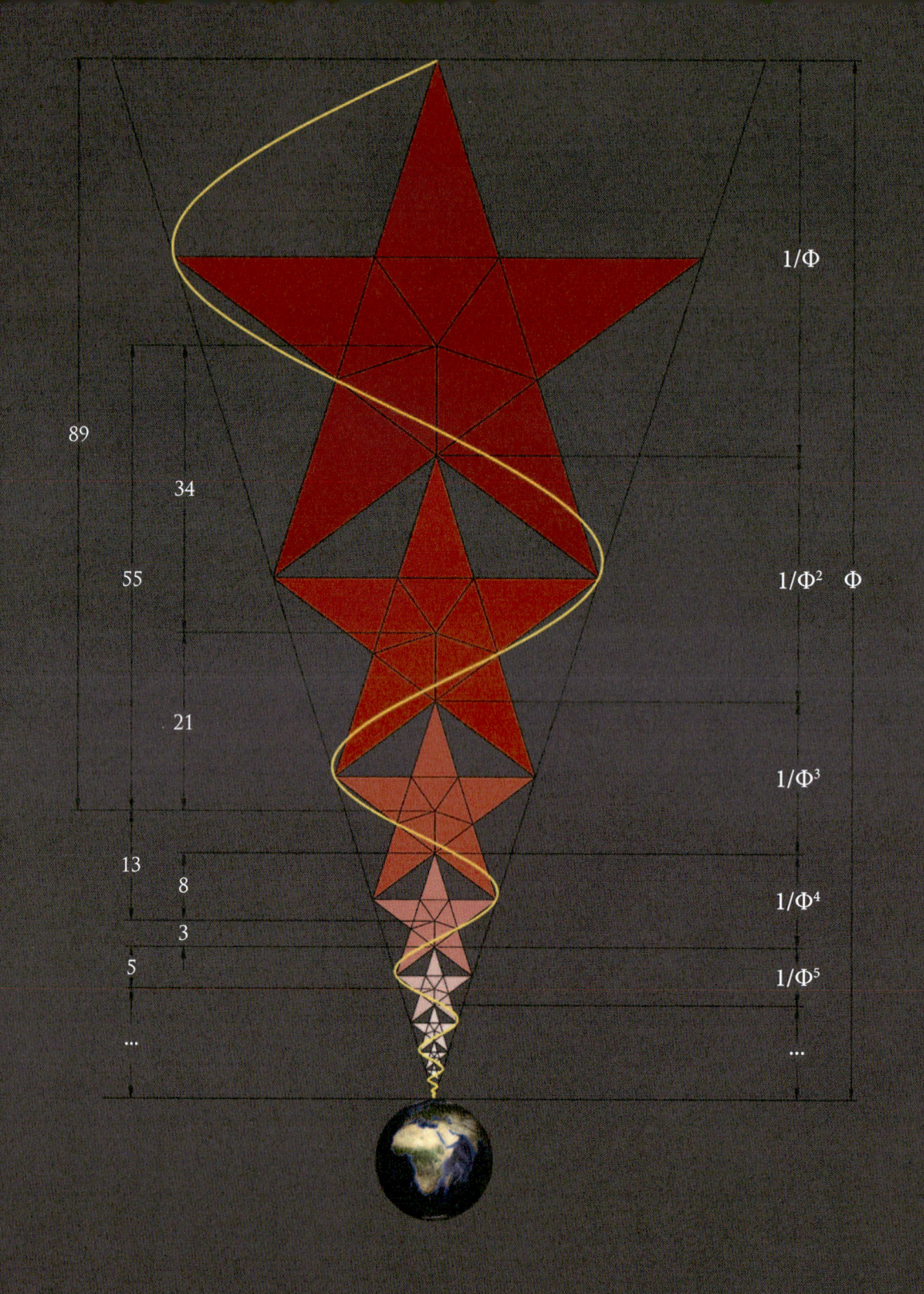

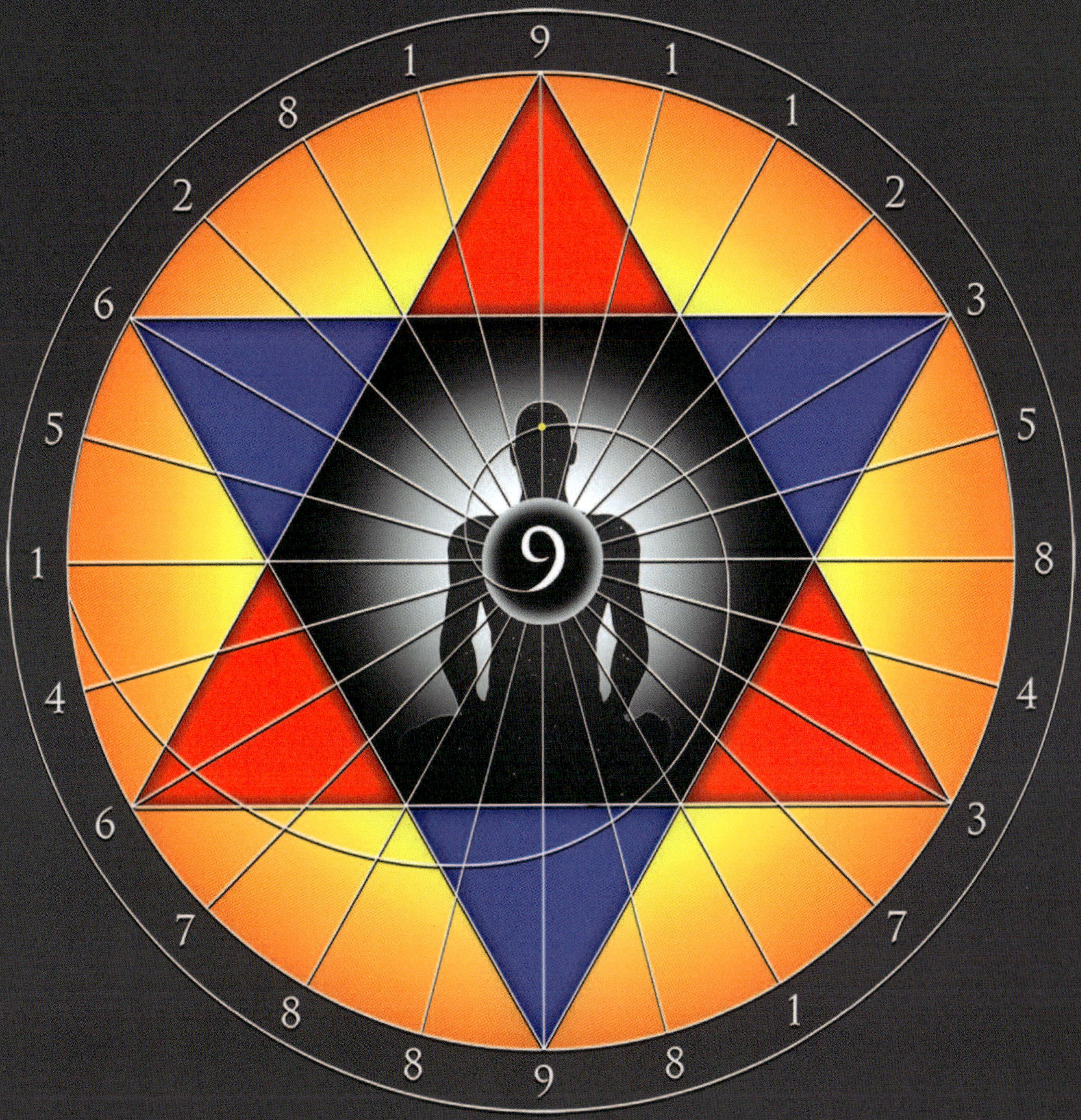
9
1
1
2
3
5
8
4
3
7
1
8
9
8
8
7
6
4
1
5
6
2
8
1
9

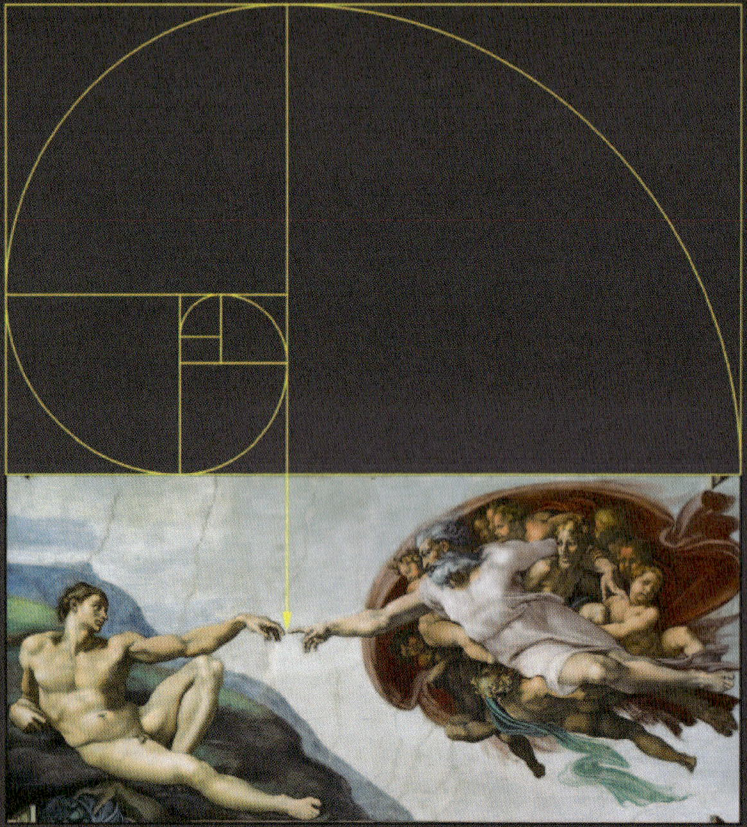

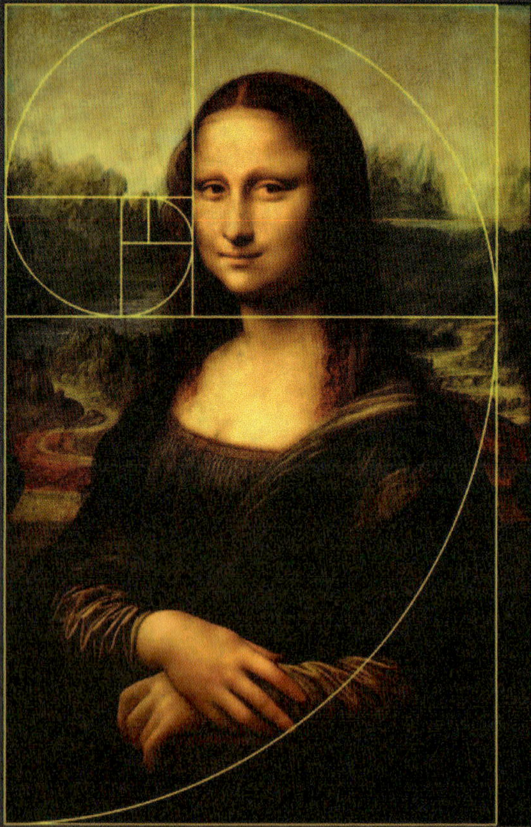

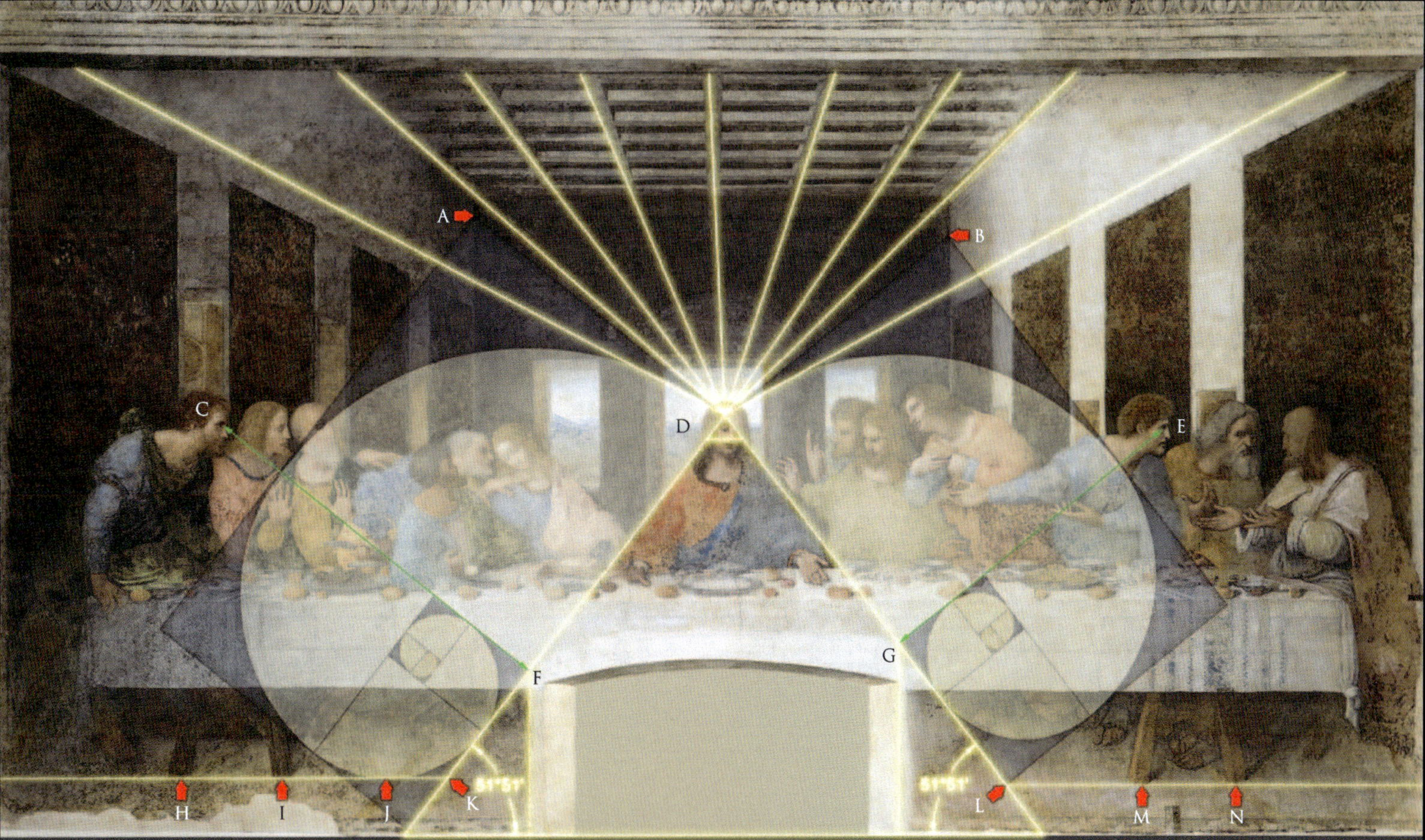
A
B
C
D
E
F
G
H
I
J
K
51°51'
51°51'
L
M
N

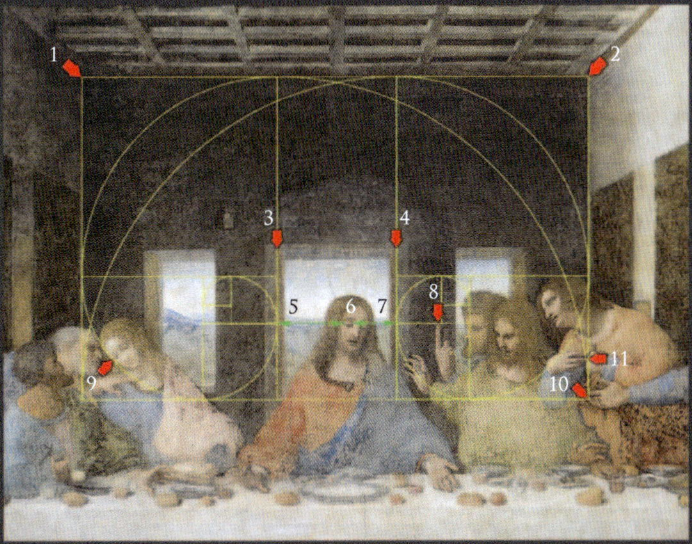
1
2
3
4
5
6
7
8
9
10
11

LEONARD G CC BY-SA 1.0
NEPTUUL CC BY-SA 3.0

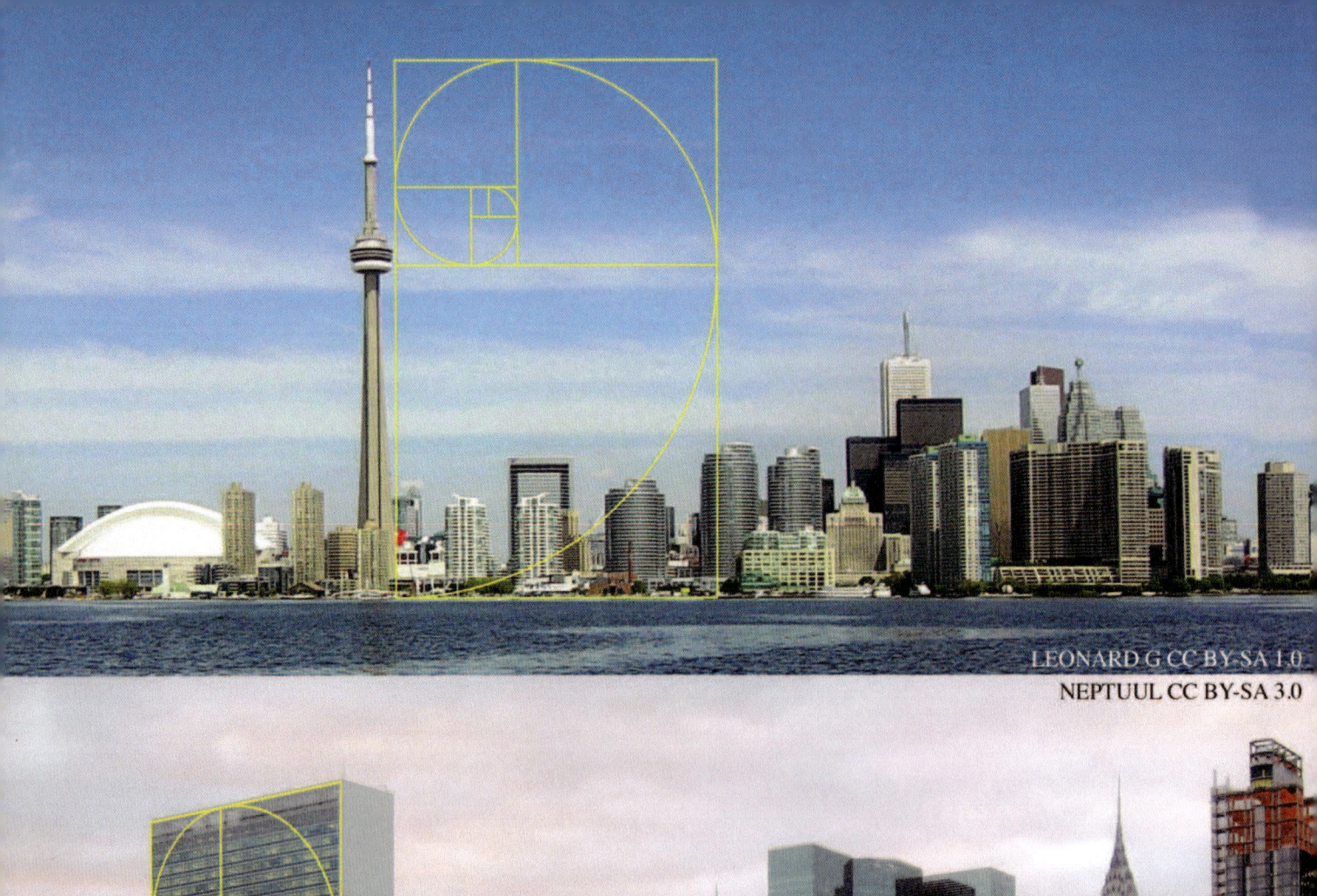

$$\frac{\alpha+\beta}{\alpha} = \frac{\alpha}{\beta} = \Phi$$

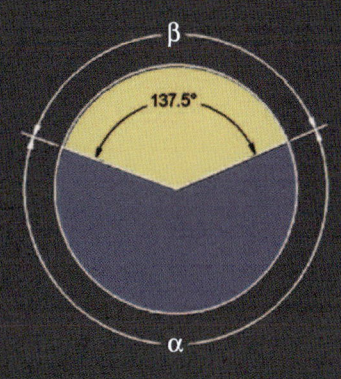

137.5°

Spiral Phyllotaxis
Divergence Angle = 137.5°

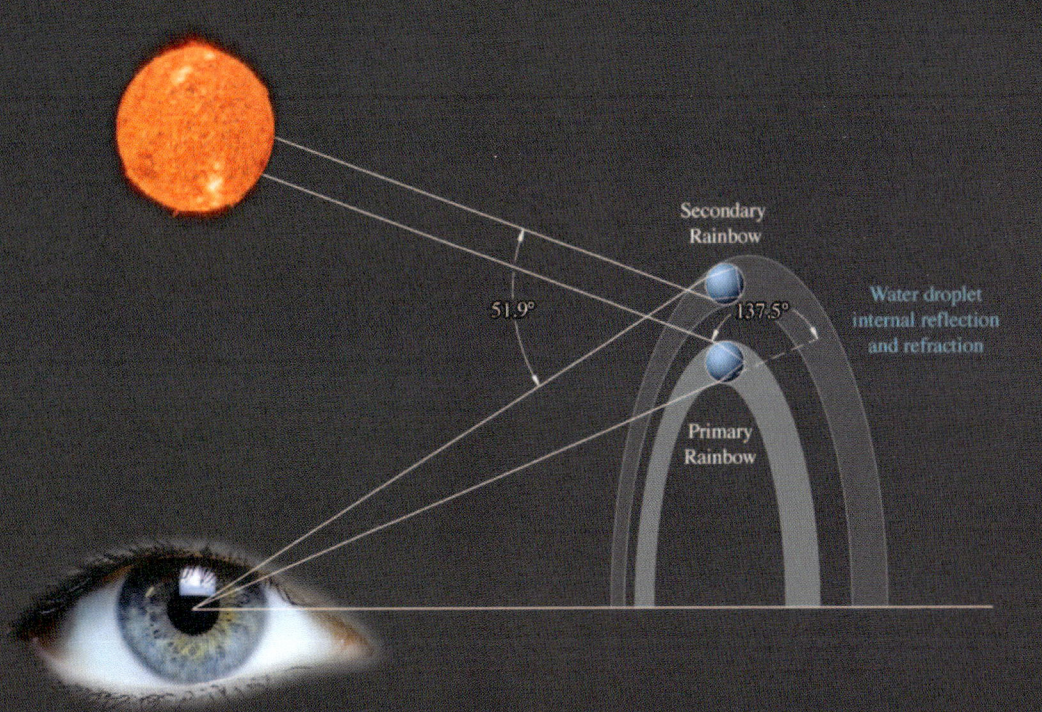

Secondary Rainbow

51.9°   137.5°   Water droplet internal reflection and refraction

Primary Rainbow

Eye photo Laitr Keiows CC BY-SA 3.0

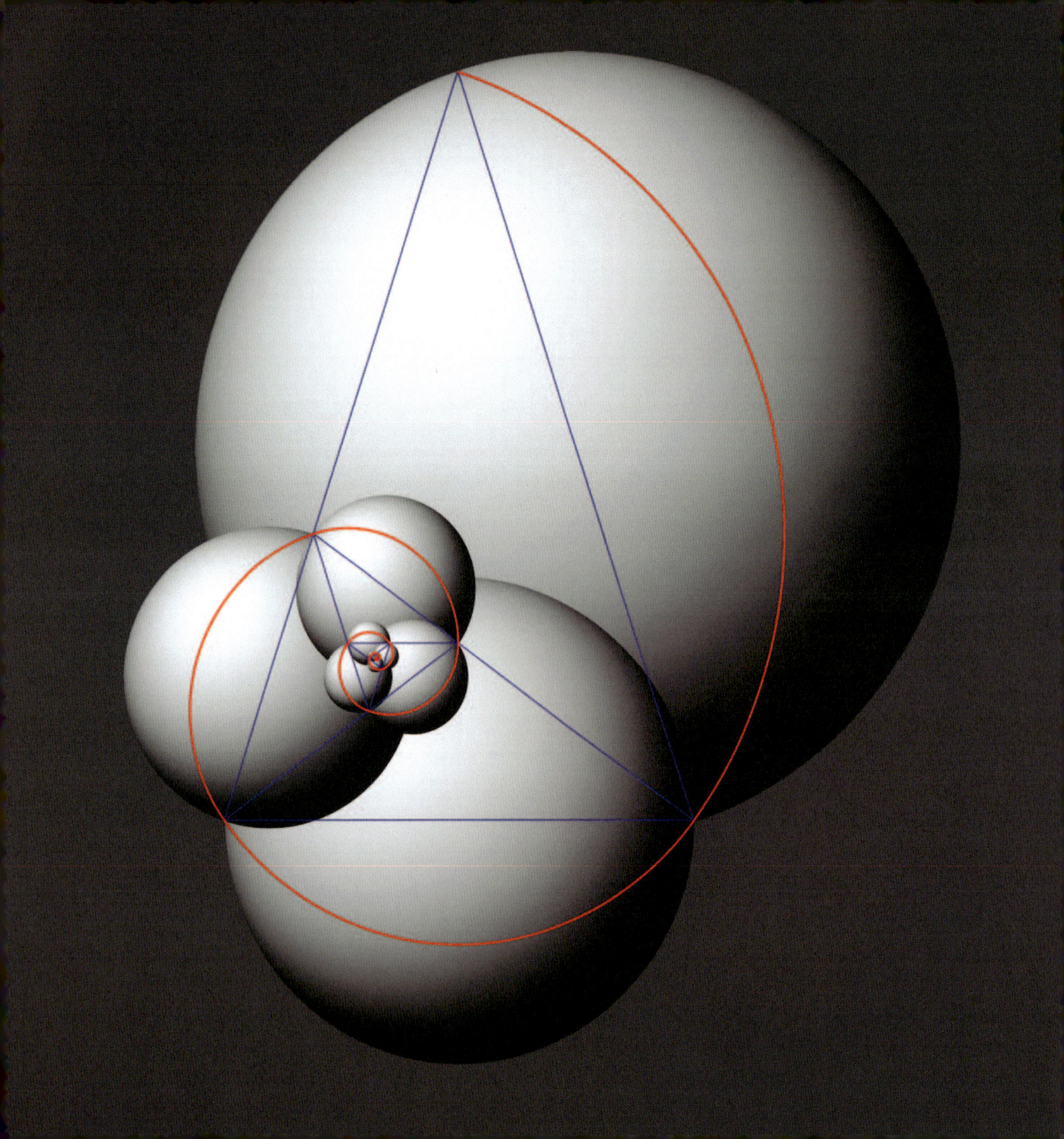

Photo by Violachick68 · CC BY-SA 3.0

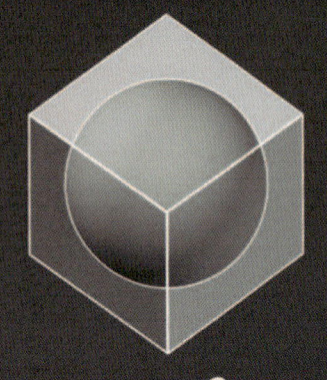

Insphere

$$\frac{V_{sphere}}{V_{cube}} \approx \pi - \Phi^2 \quad (99.99\%)$$

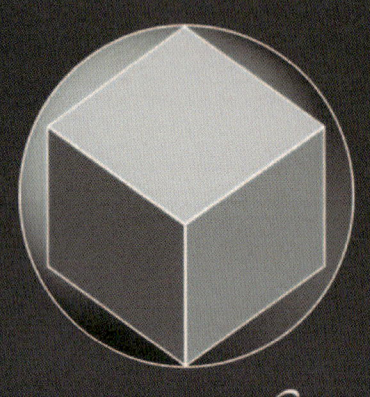

Circumsphere

$$\frac{V_{sphere}}{V_{cube}} \approx e \quad (99.9\%)$$

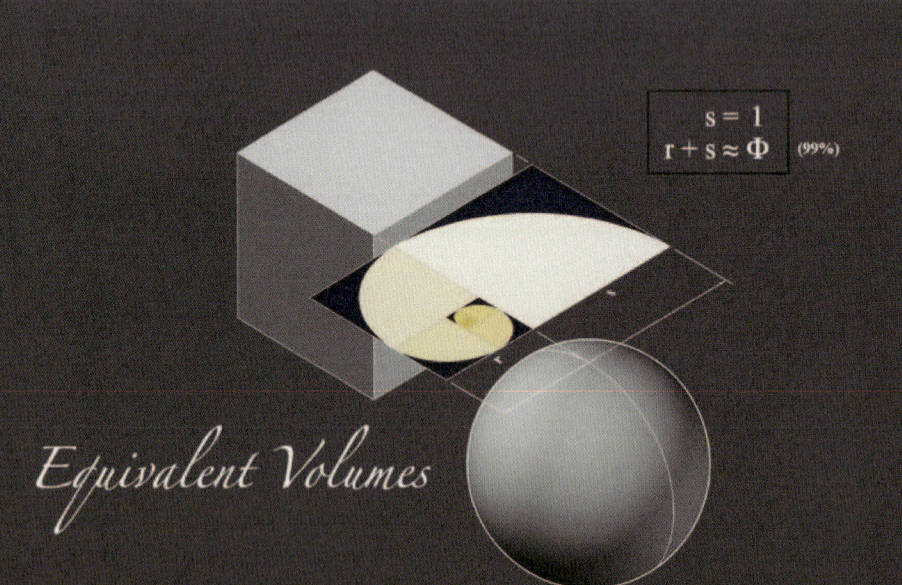

Equivalent Volumes

$$s = 1$$
$$r + s \approx \Phi \quad (99\%)$$

THE PYTHAGOREAN COMMA

TRANSCENDENTAL MATHEMATICS

ORBITAL MECHANICS

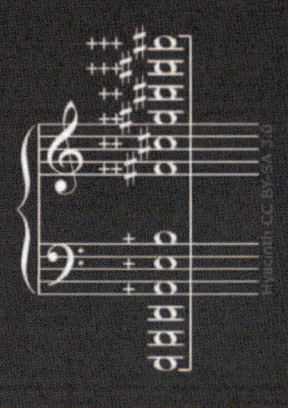

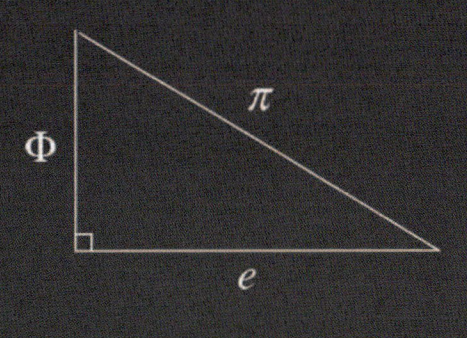

Time from Summer Solstice to the next Summer Solstice
365.2421897 days

1.618099389...   *2.718281828...*

Twelve Just Perfect Fifths

Seven Octaves

$$\frac{\left(\frac{3}{2}\right)^{12}}{\left(\frac{2}{1}\right)^{7}} \approx \frac{\Phi^2 + e^2}{\pi^2} \approx \frac{\text{Solar Year}}{360°}$$

3.141592654...   *Full Circle*

WORLD

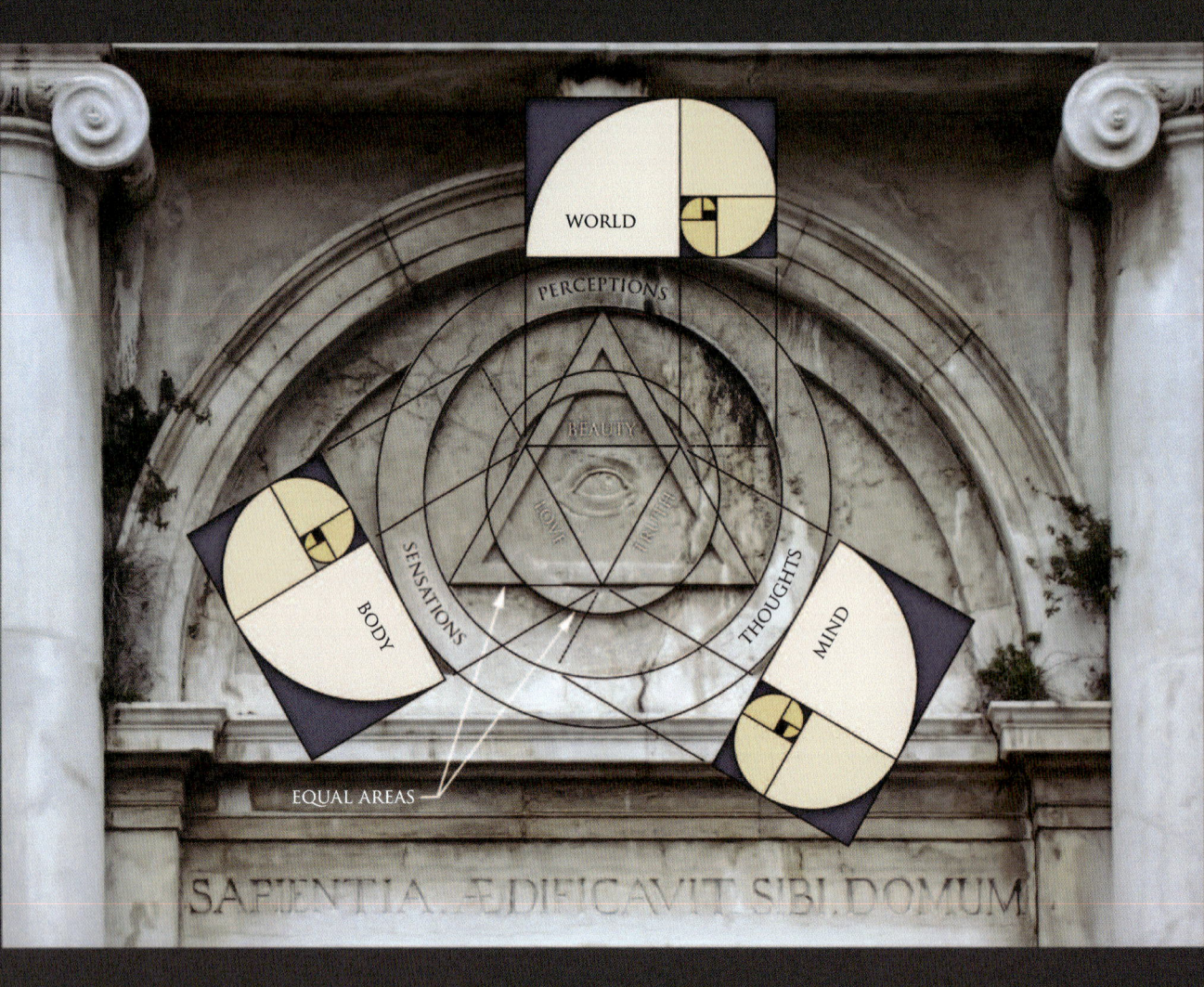

PERCEPTIONS

BEAUTY

SENSATIONS   HOME   TRUTH   THOUGHTS

BODY   MIND

EQUAL AREAS

SAPIENTIA ÆDIFICAVIT SIBI DOMUM

Equal Lengths

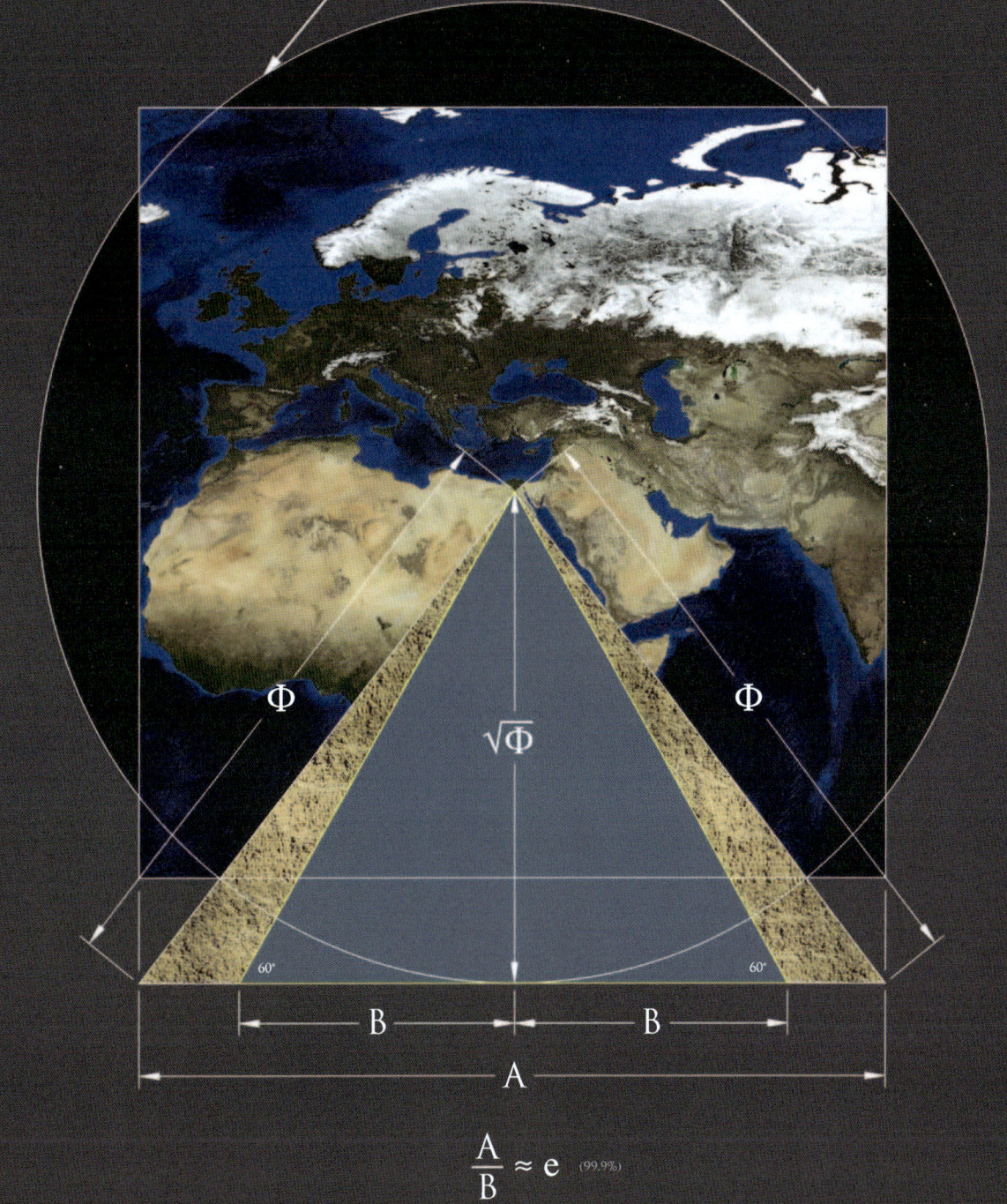

Φ √Φ Φ

60° 60°

B B

A

$$\frac{A}{B} \approx e \quad (99.9\%)$$

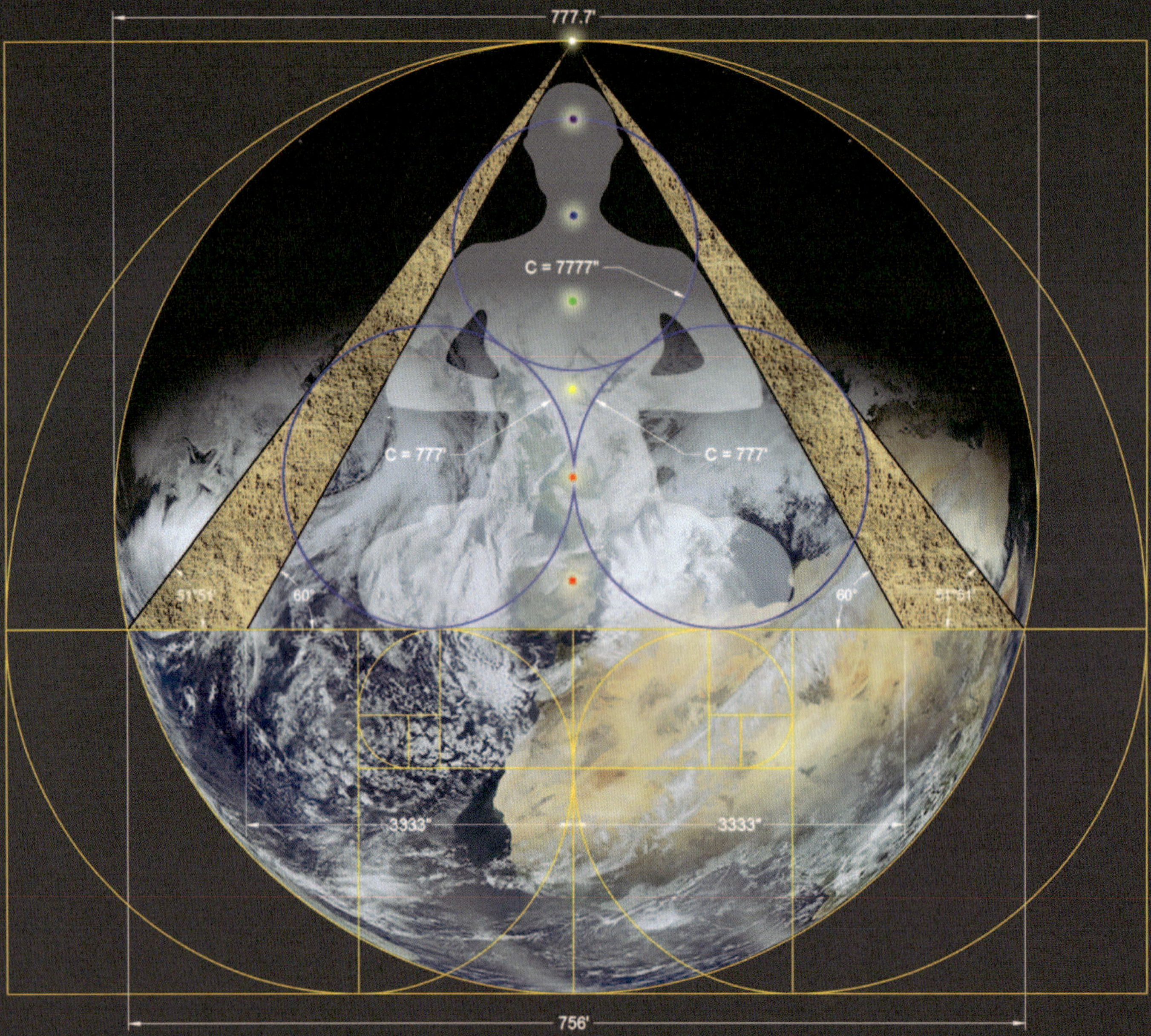
777.7'
C = 7777"
C = 777'
C = 777'
51°51'
60°
60°
51°51'
3333"
3333"
756'

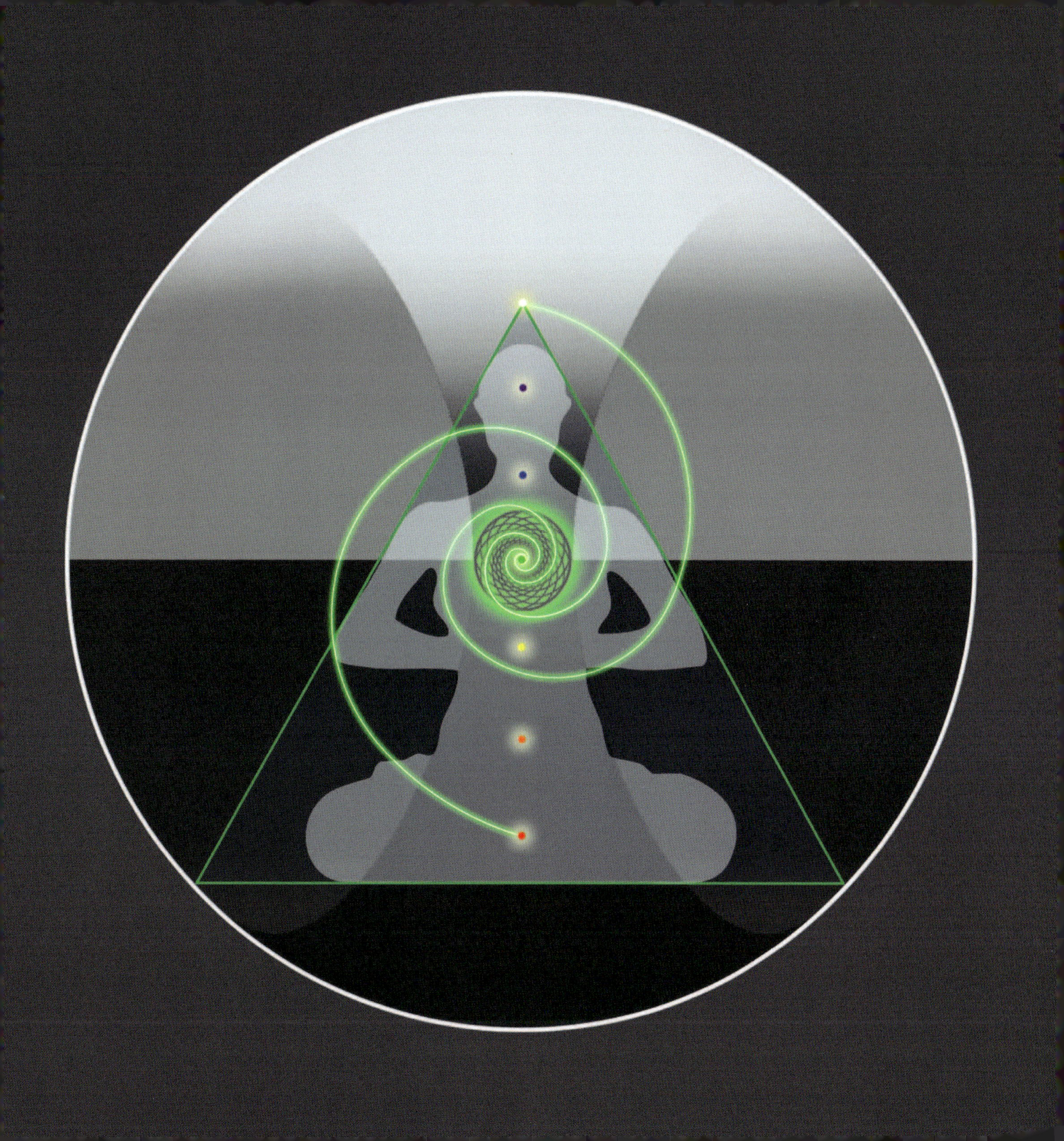

$$A+B \approx \pi$$
$$(A+B)E \approx e$$

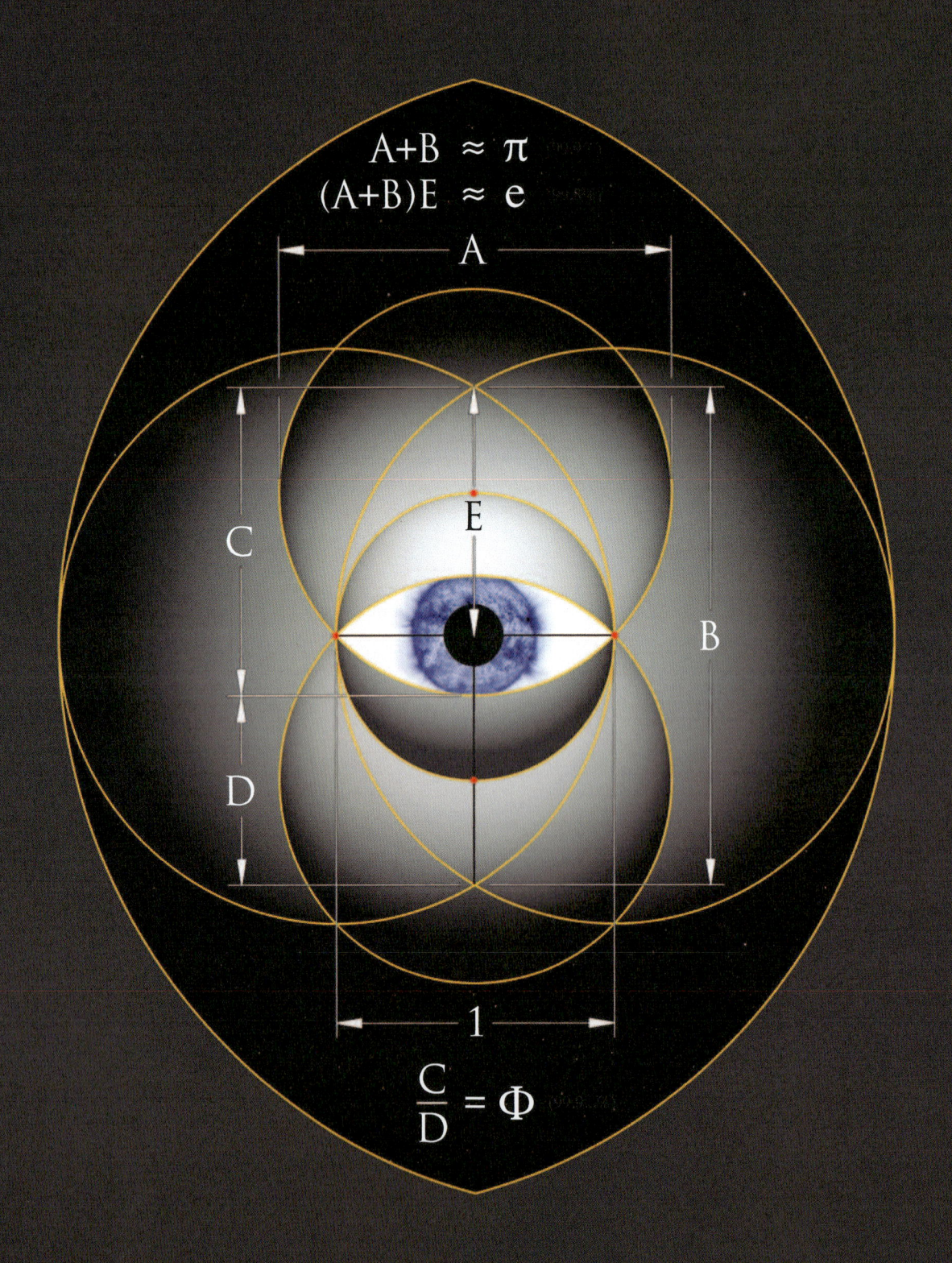

$$\frac{C}{D} = \Phi$$

Daniel Mayer CC BY-SA 4.0

.666...

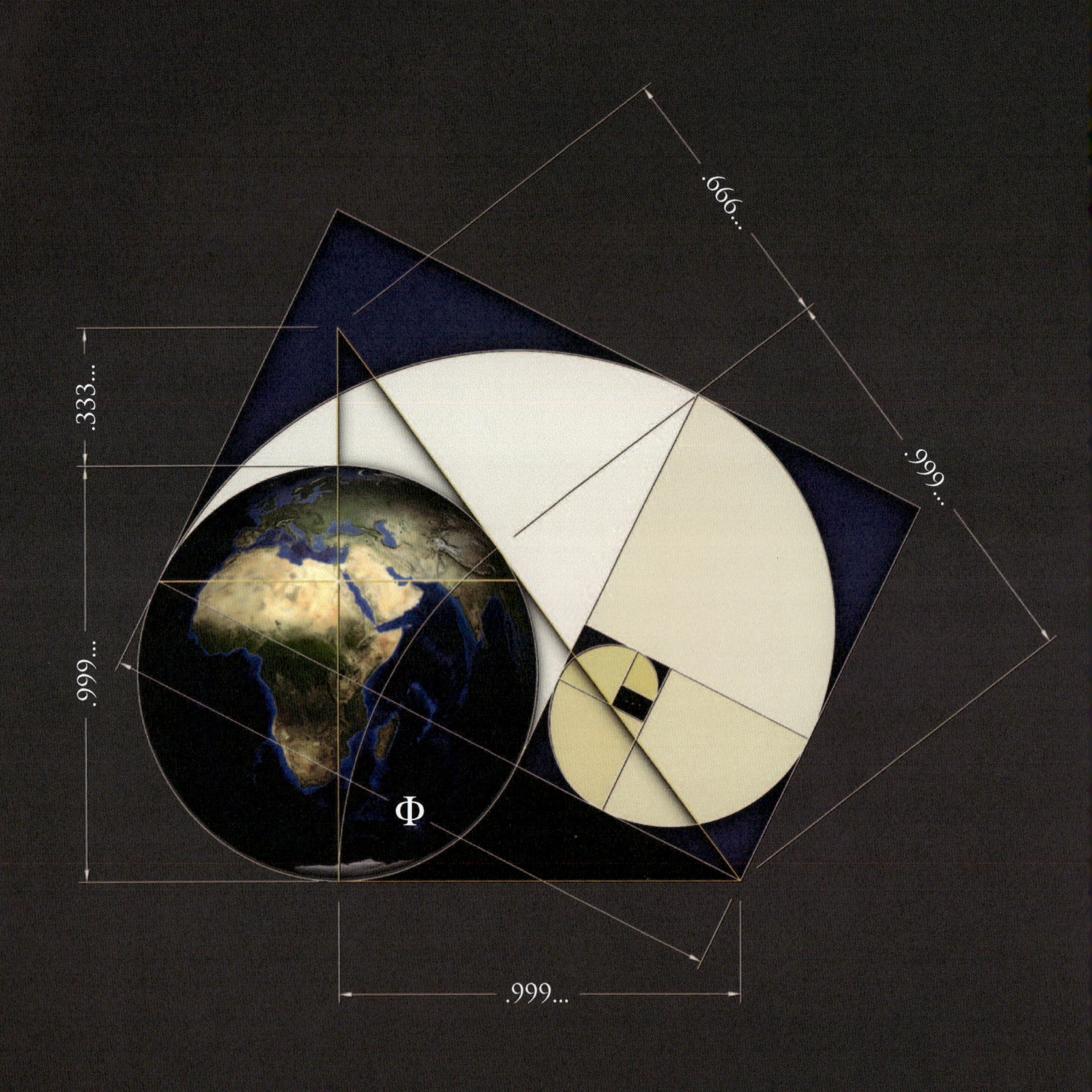

.333...

.999...

.999...

.999...

Φ

North Pole   66.6°   Ecliptic

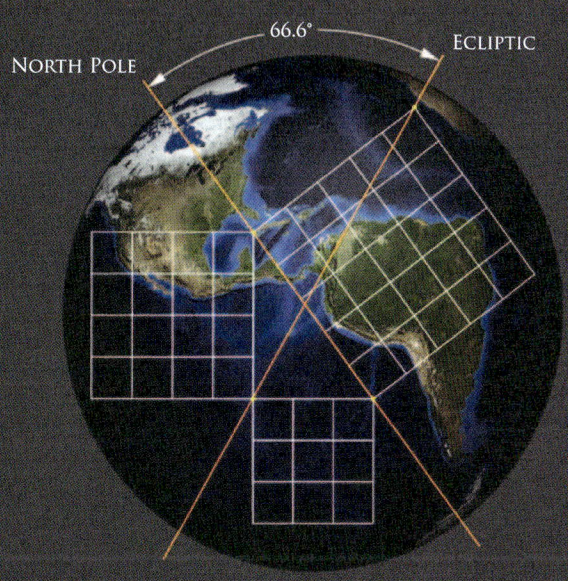

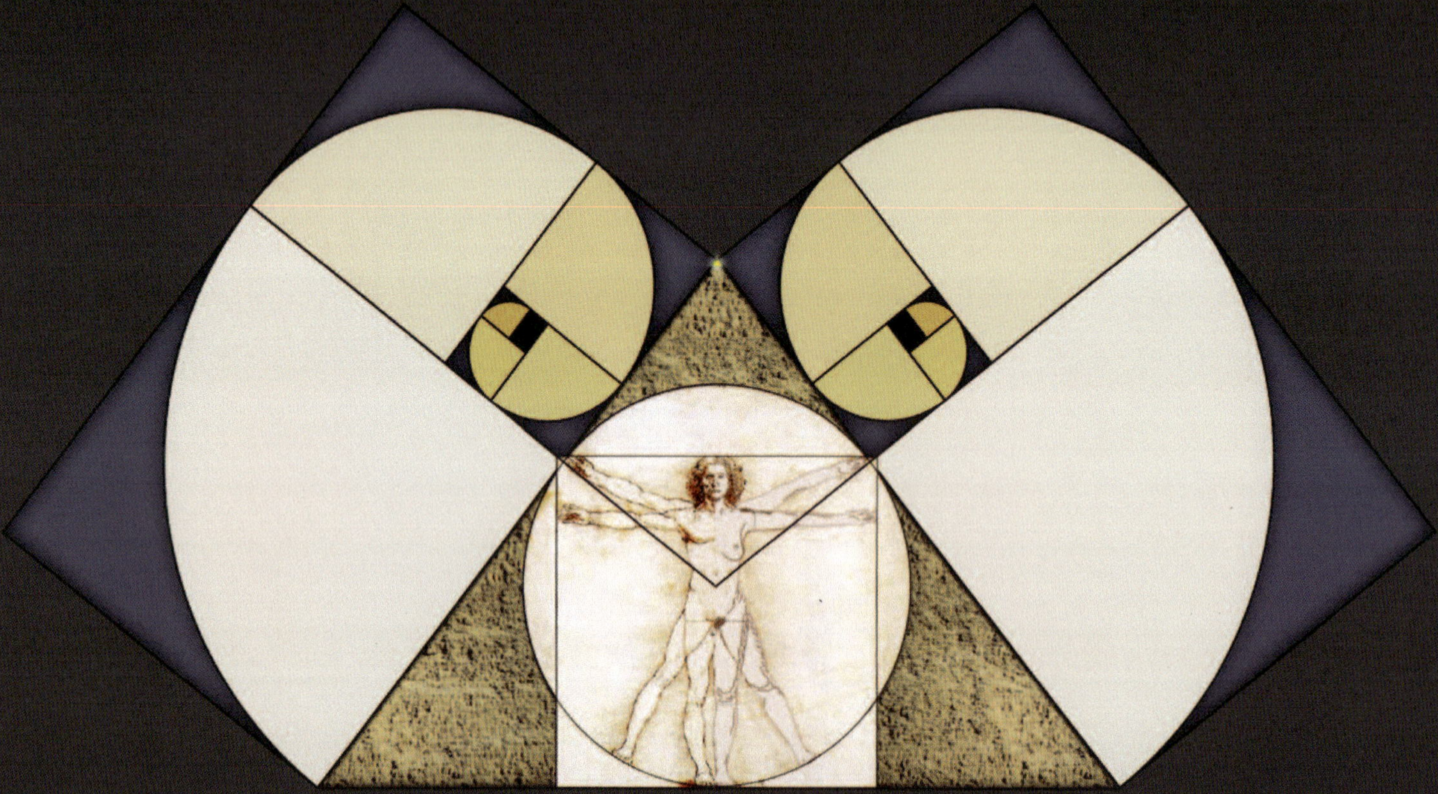

Longest Distance on Land

$$\frac{8468 \text{ Miles}}{5222 \text{ Miles}} \approx \frac{5222 \text{ Miles}}{3246 \text{ Miles}} \approx \Phi \; _{(99.9\%)}$$

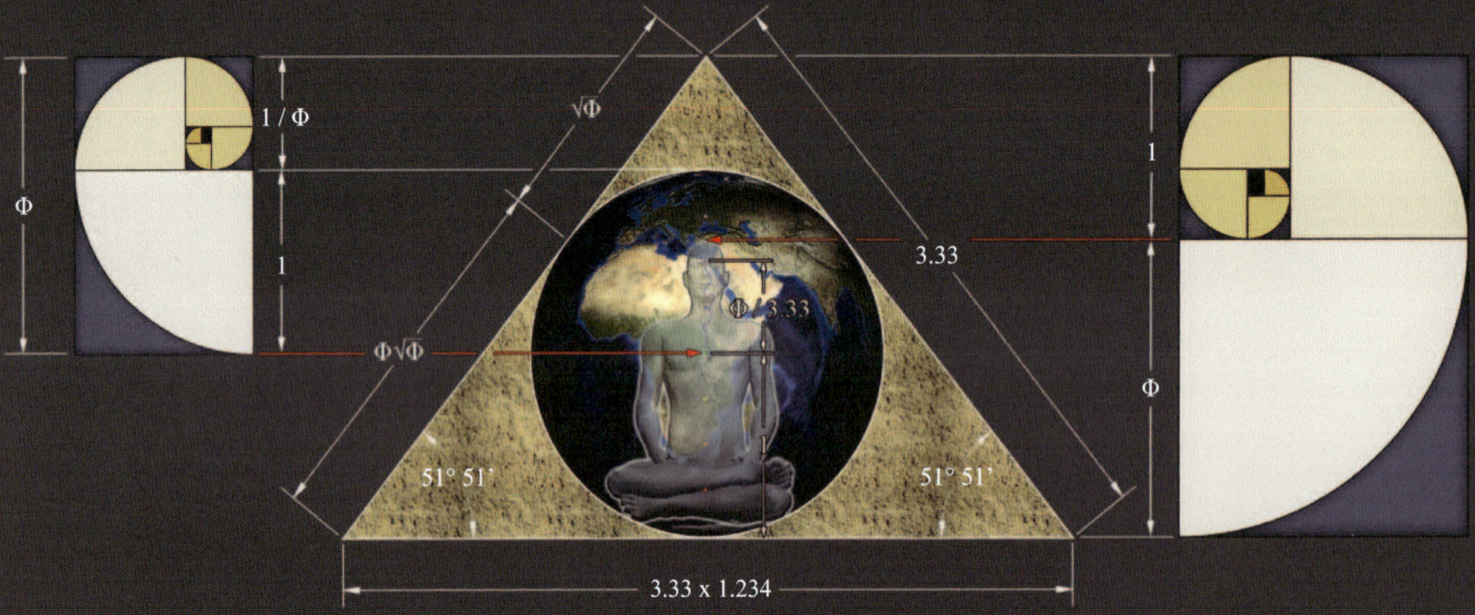
1 / Φ
√Φ
Φ
1
1
3.33
Φ / 3.33
Φ√Φ
Φ
51° 51'
51° 51'
3.33 x 1.234

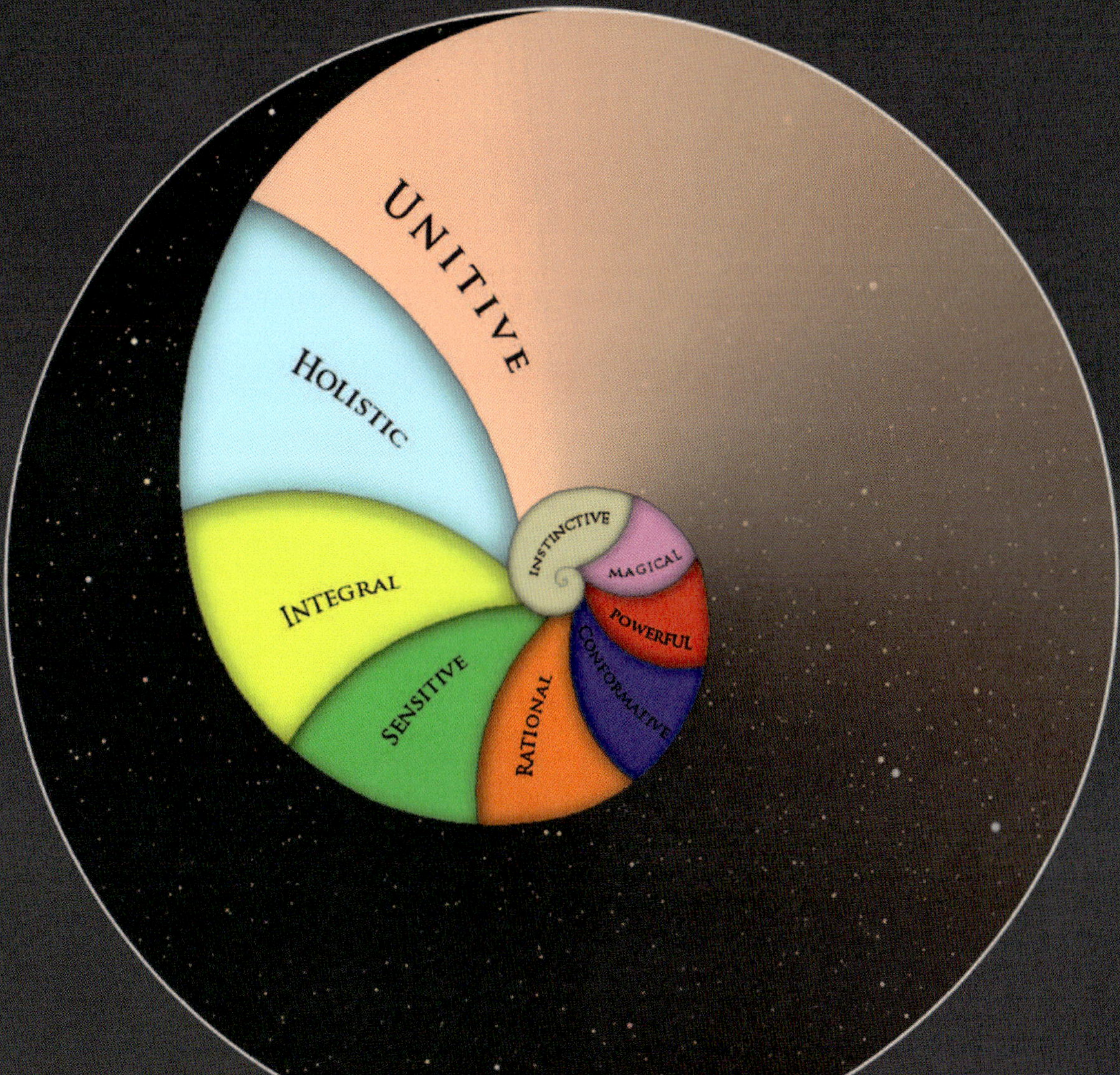
Unitive
Holistic
Integral
Sensitive
Rational
Conformative
Powerful
Magical
Instinctive

TIME

PRESENT
END OF ANTIQUITY
RISE OF CITY STATES

WRITING

THE WHEEL

DOMESTICATION OF ANIMALS

HUMANS
VERTEBRATES
ANIMALS

MULTICELLULAR LIFE

PHOTOSYNTHETIC BACTERIA

TIME

FORMATION OF THE EARTH

UNIVERSE	10^{26}	
VIRGO SUPERCLUSTER	10^{24}	
MILKY WAY GALAXY	10^{21}	
SOLAR SYSTEM	10^{13}	MACROCOSM
SUN	10^{9}	
EARTH	10^{7}	
HUMAN	1	
BACTERIUM	10^{-6}	
ATOM	10^{-10}	MICROCOSM
PROTON	10^{-16}	

SCALE

NASA / ESA

Φ   Φ

Φ²   Φ²

1   1

Φ   Φ

Φ

$\sqrt{\Phi}$

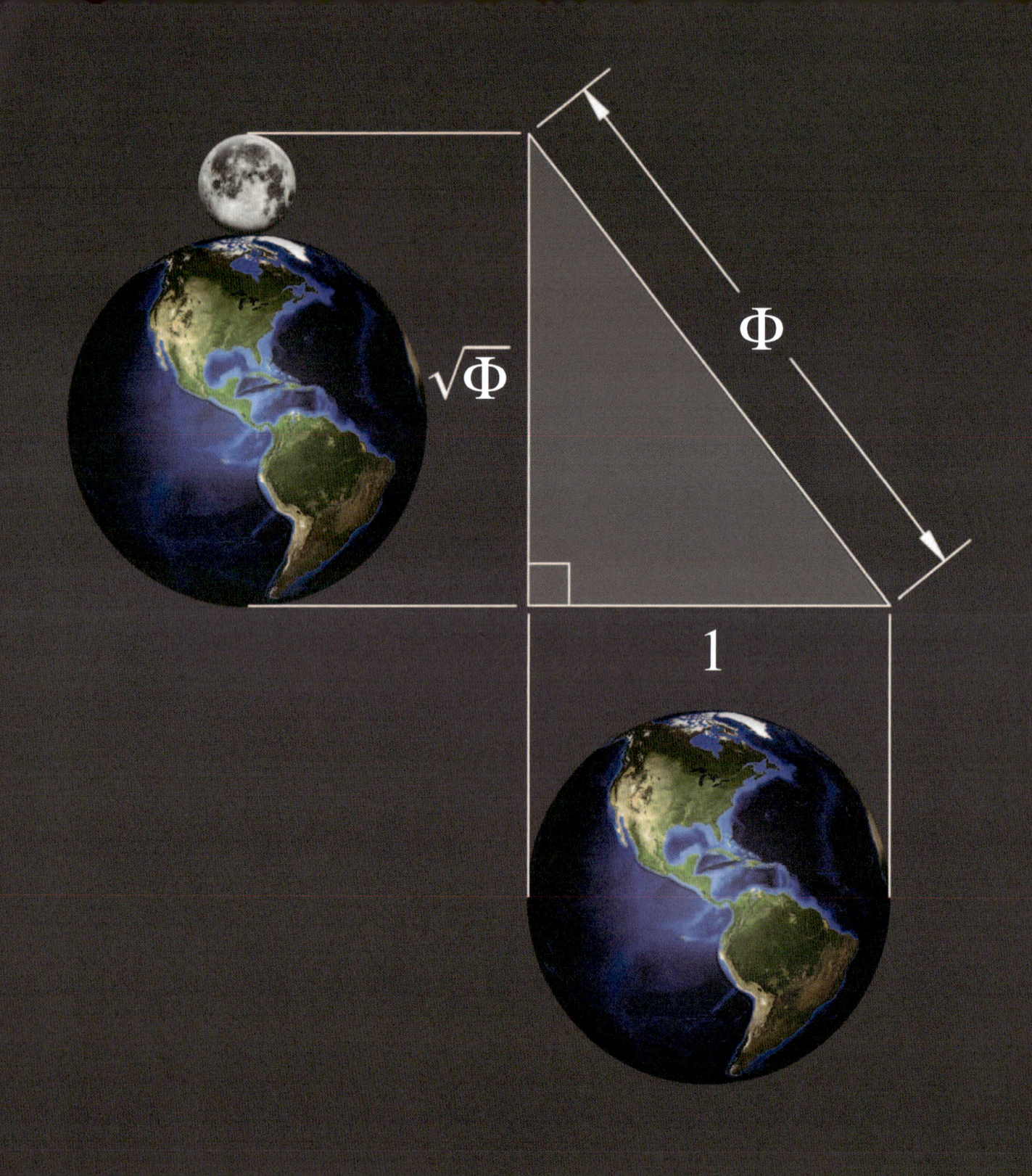

Φ

1

10000 KM
(99.98%)

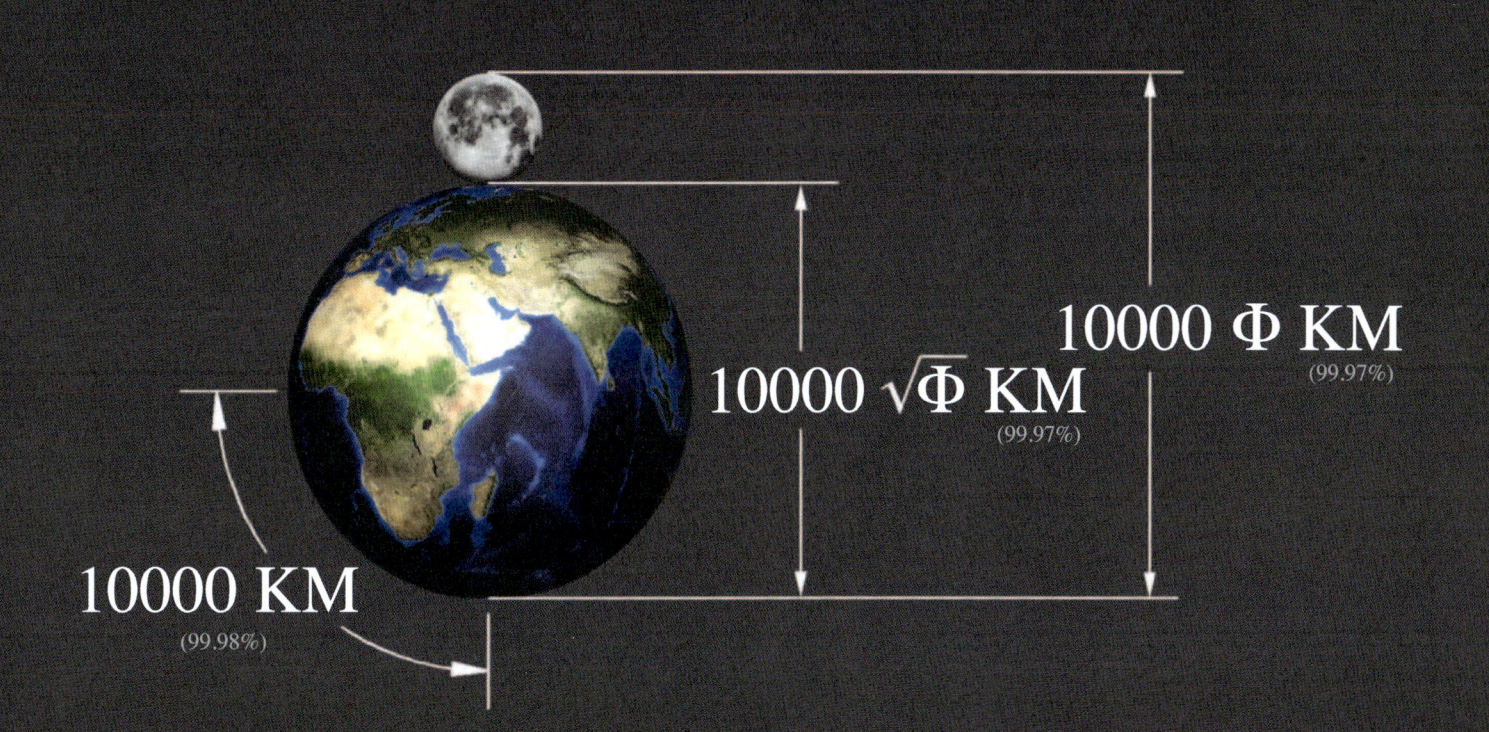

10000 √Φ KM
(99.97%)

10000 Φ KM
(99.97%)

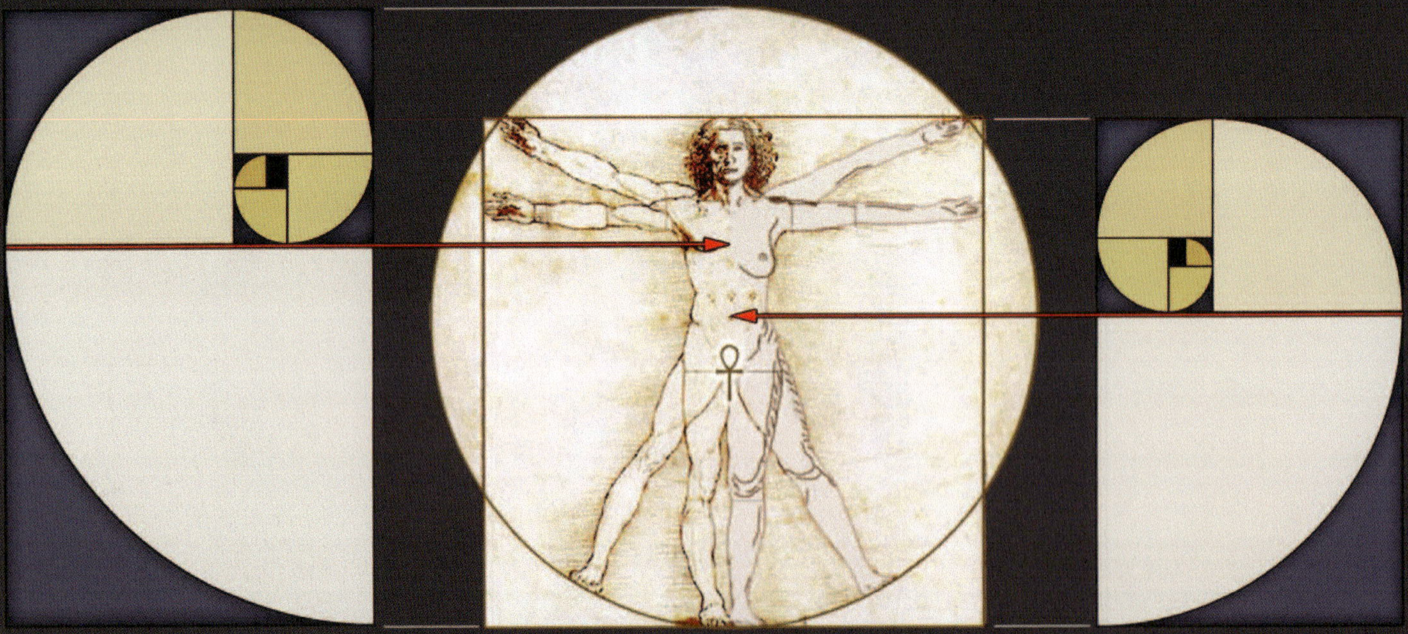

AXIAL VIEW

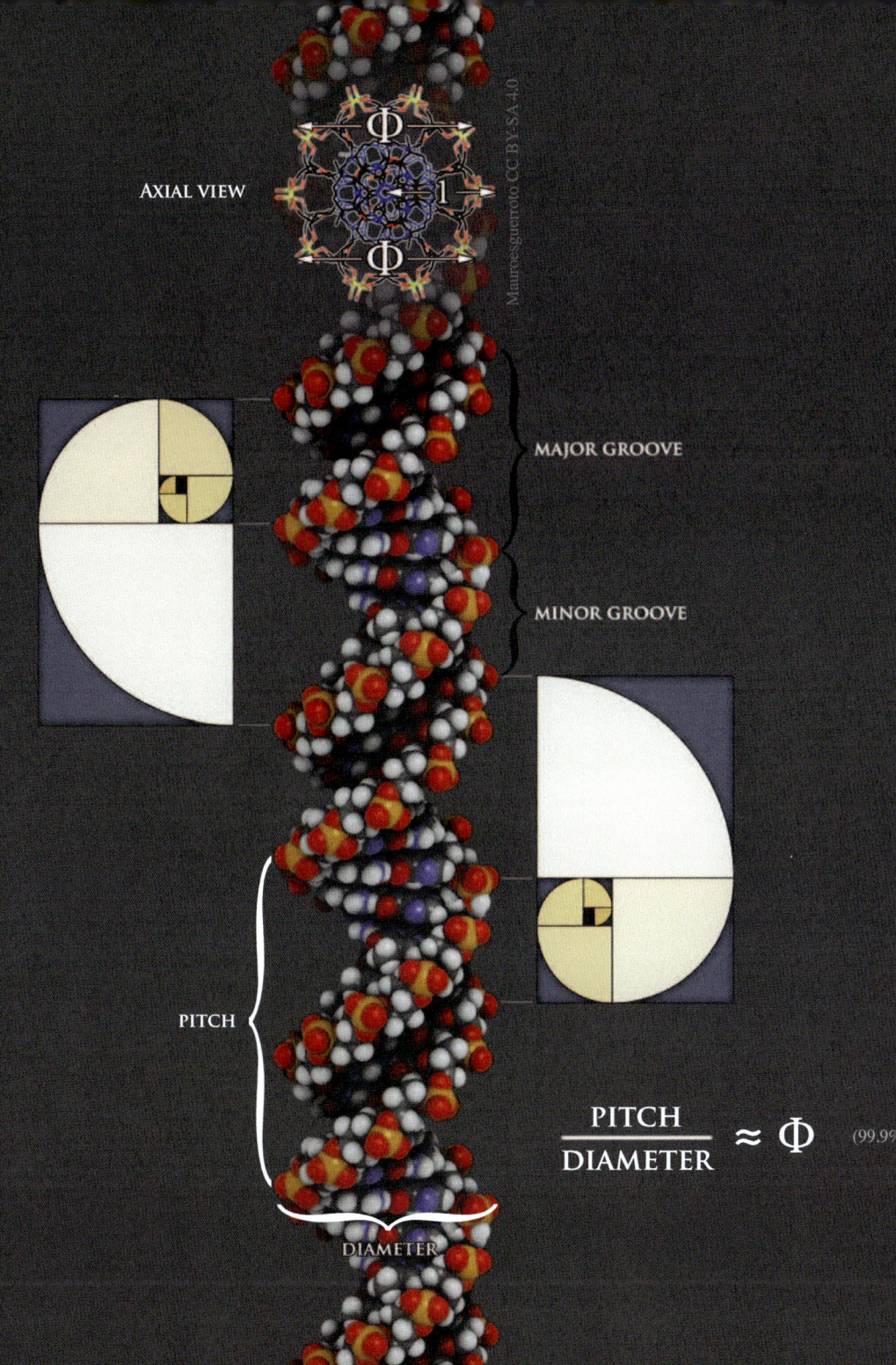

MAJOR GROOVE

MINOR GROOVE

PITCH

$$\frac{\text{PITCH}}{\text{DIAMETER}} \approx \Phi \quad (99.9\%)$$

DIAMETER

DEOXYRIBONUCLEIC ACID

Endnotes

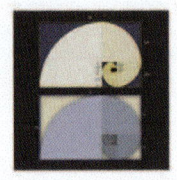

1. Golden and Fibonacci Rectangles

The golden rectangle is based on the proportion 1:Φ. When the height is 1 then the overall width is Φ, and each successive division is found by dividing the preceding term by Φ. The divisions spiral around, ever smaller, and the illustration ends with a smaller golden rectangle framing a black space. Imagine scaling down the large golden rectangle such that a copy of it would fit within the smaller golden rectangle. This scaling process could be repeated an infinite number of times such that the process of involution would eternally follow the spiral inward into the microcosm. The process could likewise be repeated in reverse such that the spiral would grow exponentially into the macrocosm with increasing powers of Φ.

The Fibonacci rectangle starts in the core as a circle. The first movement is to enclose the infinity of points equidistant from the circle's center with a square having an edge length of 1, thereby rationalizing the infinite. The next movement is to add the square's edge to the previous form, yielding a second square of edge length 1 (0+1=1). The third movement is to add the square of edge length 1 to the previous square (1+1=2). The iterative process repeats ad infinitum following the Fibonacci sequence. The physical world is built of structures at all scales and is therefore discrete—for example you can't have less than a whole seed in a sunflower and thus, we see natural growth following discrete "Fibonacci" distances that rationalize and approximate Φ.

2. Golden Pentagram

Pentagrams exhibit a self-similarity that encodes Φ in ever smaller and ever larger scales. Starting with the large red regular pentagram and going inward you see a blue inverted pentagram inscribed inside. This process is repeated and inverted again inside the blue pentagram as the smallest red pentagram. Golden rectangles are arrayed around the circumference of the circle, demonstrating how Φ is encoded in myriad self-similar divisions.

Imagine connecting the points of the largest red pentagram in a pentagon and extending its edges outwardly to form another larger inverted pentagram. This evolutionary process would result in ever-larger Φ relationships.

3. Phive

Phi is usually represented algebraically as (1+√5)/2 because this is the solution to the quadratic formula that arises out of golden ratio's definition in (α+β)/α = α/β = Φ. This is approximated in decimal form as 1.618033989… but neither of these representations capture the beauty I see in the golden ratio's self-similarity—its "as above so below" quality, if you will.

I feel that a more appropriate representation of Φ is the measure $.5 \times 5^{.5} + .5$ inscribed in a pentagon because this combined algebraic/geometric representation is composed of five fives. This simultaneous whole-brained approach captures something of the beauty of the golden ratio in how it manifests "phiveness."

4. Jacob's Ladder

The ladder Jacob saw in his apocryphal vision was a bridge between heaven and earth. The infinite rungs of any ladder between macrocosm and microcosm would necessarily be made by comparing the whole to the larger in exactly the same proportion as the larger is compared to the smaller, which is the divine proportion.

The sinuous golden path that consciousness takes on the way up or down the ladder reminds me of how successive ratios of terms in the Fibonacci sequence approximate Φ, by being alternately too large and then again too small, but always getting closer to perfection. For example, $89/55 \approx 1.61818$ and $144/89 \approx 1.61798$, while $\Phi \approx 1.61803$. Perhaps the stone Jacob rested on was a metaphor for the Earth itself.

5. Fibonacci Sequence (mod 9)

We already have lots of experience with modular arithmetic, but most are probably not aware of it. For example, we know that if it was 12 noon and 21 hours had elapsed, it would be 9 the next morning. This is written as $12 + 21 \equiv 9 \pmod{12}$. In modulo arithmetic we always reset the count when the numbers wrap around past the modulus, which is 12 in this example.

In much the same way, in Mod 9 we start the clock over when the numbers wrap around past 9, so $10 \equiv 1 \pmod 9$. Mod 9 is actually far easier to figure out than Mod 12 because it is congruent with digit sum arithmetic. In other words, to find any Mod 9 answer you simply add up the number's digits. For example, the digit sum of the number 89 is $8 + 9 = 17$. All Mod 9 answers are single digits so in this example we must add the digits again, as $1 + 7 = 8$. This is written $89 \equiv 8 \pmod 9$. The result is known as the number quality, essence, or digital root. Another way of saying this is "the digital root of 89 is 8."

The digital roots of the Fibonacci sequence repeat with a period of 24. In the illustration I arrayed the 24 digital roots of the Fibonacci sequence clockwise around the circle.

Each one of these essential numbers has a diametrical opposite, 180° away. Pick any number and follow its straight path through the heart of the figure to the opposite side and observe that all opposite number pairs $\equiv 9 \pmod 9$. The numeral 9 in the center spirals out from the microcosm of the heart, into the vast macrocosm as the Fibonacci sequence spins without bound around the outer wheel.

Term Number	Fibonacci Term	Digital Root	Term Number	Fibonacci Term	Digital Root
1	0		25	46368	9
2	1	1	26	75025	1
3	1	1	27	121393	1
4	2	2	28	196418	2
5	3	3	29	317811	3
6	5	5	30	514229	5
7	8	8	31	832040	8
8	13	4	32	1346269	4
9	21	3	33	2178309	3
10	34	7	34	3524578	7
11	55	1	35	5702887	1
12	89	8	36	9227465	8
13	144	9	37	14930352	9
14	233	8	38	24157817	8
15	377	8	39	39088169	8
16	610	7	40	63245986	7
17	987	6	41	102334155	6
18	1597	4	42	165580141	4
19	2584	1	43	267914296	1
20	4181	5	44	433494437	5
21	6765	6	45	701408733	6
22	10946	2	46	1134903170	2
23	17711	8	47	1836311903	8
24	28657	1	48	2971215073	1

6. Masterpieces

Perhaps the divine proportion is the hidden reason these works of art became two of the most iconic masterpieces in the Western tradition. You don't have to consciously understand with the left hemisphere of your brain that these works are based on the divine proportion to have your right brain know and heart feel their beauty. The divine proportion is part of you.

In a paper entitled "More than a neuroanatomical representation in The Creation of Adam by Michelangelo Buonarroti, a representation of the Golden Ratio," (Clinical Anatomy 17 July 2015), the co-authors Deivis De Campos, Tais Malysz, João Antonio Bonatto-Costa, Geraldo Pereira Jotz, Lino Pinto De Oliveira Junior, and Andrea Oxley da Rocha build on Meshberger's 1990 discovery that, "The Creation of Adam represented in the Sistine Chapel shows God surrounded by a drape that has the shape of what he believed to be the sagittal section of a human brain." (http://bit.ly/1UQdObR)

The co-authors demonstrate that not only is the famous gap between God's and Adam's fingers at the golden ratio point in The Creation of Adam fresco, but the location of this point within the overall length of Sistine Chapel ceiling is likewise divinely proportioned.

I illustrated just the first part of this harmony using a golden rectangle. The gap where Adam receives the divine spark is precisely and fittingly pointed to using the divine proportion.

7. Ph-eyes

In Ph-eyes I show how Leonardo da Vinci's "The Last Supper" is based on the Great Pyramid's slope angle and golden rectangles. I drew lines sloped at 51°51' through the corners of the doorframe at F and G. These lines intersect in the center of Jesus' head, just above his right eye. This eye is framed at D, which is located on the centerline of his body and the centerline of the entire composition. Seen in the context of the pyramid, this eye functions as the iconic *Eye of God* within the pyramidion, a symbol echoed centuries later as the "eye of providence" on the US dollar bill.

The lines of perspective on the ceiling radiate from the top of the pyramid much like rays of Sun on the horizon. I next drew two horizontal lines along the base of the stools using pairs of points H, I and M, N respectively (the resulting horizontal lines are at slightly different levels).

Placing golden rectangles along the pyramid's faces, both anchored at the top of the pyramid, I scaled the left rectangle so that it would be anchored tangent to the arc at J and anchored the right rectangle's corner at L. Both golden rectangles are also anchored at A and B along the vertical edges of the back wall.

Verification that Leonardo used this setup comes from the eyes precisely pointed out at C and E with the green arrows that come from the original anchors at F and G, respectively. I titled my illustration Ph-eyes as a concatenation of the Greek letter "phi" and "eyes," which Leonardo subconsciously draws our attention to using the divine proportion.

The detail at the bottom has two mirrored golden rectangles anchored at the corners of the back wall at 1 and 2. The primary divisions of the divine proportion parallel the wooden doorframes at 3 and 4. A transverse division at 5 and 7 establishes a line that passes through the center of Jesus' eye at 6. Points 8, 10 and 11 are located at the tips of various disciples' fingers and point 9 is tangent to the face of the feminine figure at Jesus' right arm.

8. Golden Architecture

The CN Tower in Toronto is the tallest structure in the Western hemisphere at 553.33 meters in height and was the tallest structure in the world when built in 1976. The glass floor and outdoor observation deck in the Main Pod are 341.99 meters (1122 feet) in height. 553.33 / 341.99 ≈ Φ (99.99%).

Le Corbusier, the lead architect of the United Nations Secretariat building, previously developed a proportional system used for building design called the Modulor, which was based on human measurements, the divine proportion, and Fibonacci numbers. (http://bit.ly/1JdKQu7)

The following link to an article details how the divine proportion was used in the Secretariat building's overall form, windows, entrance and floor plan: http://bit.ly/1K3cmzL

9. Golden Angle

Applying the divine proportion to a circle results in what is called the golden angle. If you divide a full circle by the golden ratio you have 360° / Φ = 222.5°, and that angle is represented by the larger blue section of the pie. The smaller section of pie is 137.5° because it is the remainder of the circle (360° - 222.5° = 137.5°). Most will find the smaller measure more practical to work with because it is less than half of a circle.

Phyllotaxis is a subject that has long fascinated scientists, the study of plant patterning based on repeated biological units. Irving Adler wrote phyllotaxis papers for over two decades, collected in Solving the Riddle of Phyllotaxis: Why the Fibonacci Numbers and the Golden Ratio Occur on Plants. (http://bit.ly/1KsO91s)

In The Algorithmic Beauty of Plants by Przemyslaw Prusinkiewicz and Aristid Lindenmayer, the authors show that phyllotatic patterning is extremely precise such that the 137.5° angle occurs in nature with an accuracy of 1 part in 1869. (http://bit.ly/1NfzDjG, page 101)

The physical optics of rainbows has also long fascinated scientists. René Descartes discovered that the primary rainbow we see is caused by a single internal reflection inside water droplets whereas the secondary rainbow is caused by a double internal reflection. The reflection angles interest me, as the primary rainbow angle is 137.5° (often referred to in the literature with its supplementary angle of 180° - 137.5° = 42.5°).

The secondary rainbow's double internal reflection results in an angle of 51.9°, and I noticed that is equivalent to the Great Pyramid's slope angle of 51° 51'. Turn to illustrations 15 and 16 to see how the Great Pyramid encodes the divine proportion.

10. Golden Triangle

There are two ways to create a spiral based on the divine proportion, one with the golden rectangle (shown in plate 1), and the other with the golden triangle (shown here in blue). The golden triangle has the same proportions as one arm of a regular pentagram. As the triangles spiral inward they each cut the edge of the larger triangle they are nested inside of at the divine proportion. The red spiral is drawn with circular arc segments centered each time on the Φ division on the opposite triangle edge. The spheres are scaled in a series of descending golden ratios. This process could continue infinitely inward and/or outward.

11. Golden Violin

In an article published in the journal Notes on Number Theory and Discrete Mathematics, Robert van Gend showed how the violin's design is based on the divine proportion in his article, "The Fibonacci Sequence and the Golden Ratio in Music" (http://bit.ly/1ij3rzn). I added another such relationship with the bottom golden rectangle that illustrates how the bridge is divinely proportioned with respect to the instrument's width. The violin depicted in my illustration is the Lady Blunt Stradivarius violin of 1721, which broke the record price for a musical instrument when it sold in 2011. Could it be that part of the violin's subconsciously perceived value comes from the fact that the instrument's design is based on the divine proportion?

12. Cubing the Sphere

The cube's relationship with the sphere encodes three important mathematical constants: e, π, and Φ. An *insphere* is a sphere inscribed perfectly so that it is tangent within a cube whereas a *circumsphere* is a sphere that fits perfectly around a cube so that the cube's corners are tangent with the sphere. I discovered that the ratio of volumes in these arrangements encodes the three constants as shown. Φ is the divine proportion, π is the ratio of a circle's circumference to its diameter, and e is the base of the natural logarithm.

Another way of comparing a cube and a sphere is by attempting to make their volumes commensurate. The divine proportion is the key to "cubing the sphere." When commensurate r, the radius of the sphere, is scaled to $1/\Phi$ while the cube's edge length s is scaled to 1, such that $1/\Phi + 1 = \Phi$. Each of these comparisons have accuracy percentages shown in parentheses.

Ferdinand von Lindemann proved in 1882 that it is impossible to square the circle exactly, and by deduction it is impossible to cube the sphere exactly. This is so because one can never truly rationalize the transcendental quality of π. Countless generations of philosophers, mathematicians and geometers have sought to square the circle and cube the sphere in heroic attempts to understand and in some way attempt to explain the ineffable. The modern knowledge of the impossibility of this perfection in the real world doesn't stop one from trying to get as close to it as possible, and the divine proportion provides a bridge from the real to the ideal forms from which the universe is constructed.

13. The Gap

The *just perfect fifth* is the most consonant interval in music, save of course for unison and the octave. *Just tuning* relates musical intervals with pure frequency ratios of small integers (the fifth is 3:2). Twelve just perfect fifths and seven octaves seem like they should be exactly the same interval, but in fact there is a small gap known as the Pythagorean Comma revealing these intervals are very slightly out of tune.

The ratio of twelve fifths to seven octaves "should be" exactly 1 but the actual value is approximately 1.014. Equal division of the octave, which became standard only in the early 19th century, eliminated this gap. In order to equalize musical intervals, the twelfth root of two was mathematically applied to all musical intervals rather than ratios of small integers (such as 3:2 or 4:3) and as a consequence, all notes now are equidistant but sound slightly out of tune. The advantage is that different instruments play much more easily together because they can all be tuned with this system, which is called Equal Temperament.

The transcendental constants e, π, and Φ can be nearly brought into geometric harmony with each other in a simple Pythagorean right triangle. This harmony is imperfect because there is a slight gap, barely visible in the illustration at the top of the triangle. Using the famous Pythagorean theorem, one can compare the squares of the triangle's sides with its hypotenuse, and the ratio is approximately 1.014.

The Earth revolves around the Sun basically 1° per day, but the correlation between the Solar Year and the full circle is inexact. The ratio of the Solar Year to the full circle of 360° is also approximately 1.014. Incidentally, the ratio of males to females worldwide is 1.014.[1]

I noticed that the gap in each of these phenomena is 99.9% the same, such that I suspect this numerical gap is an essential clue in the Great Mystery. This clue beautifully entangles and spans the entire Quadrivium, whose traditional subjects include math, music, geometry and astronomy. In other words the Quadrivium expresses number in itself, number in time, number in space, and number in space-time. This totality recapitulates the creation of the universe by, of and in awareness through the extension of number.

---

1. See People and Society::WORLD / Sex Ratio at https://www.cia.gov/library/publications/the-world-factbook/geos/xx.html.

14. Phi of God

I took this photo of the Eye of God symbol over the entrance of the Magdalene round church in Venice, Italy and superimposed golden rectangles to reveal the harmony in this fundamental geometry. The triangle design symbolizes the trinity, which in the Christian tradition is represented as the Father, the Son and the Holy Spirit. All three are inseparable yet irreducible aspects of divinity. A far older tradition expressed this trinity as Osiris, Horus and Isis.

I intuit that all such trinities ultimately symbolize the triple nature of consciousness we experience in our thoughts, sensations and perceptions. We feel sensations and emotions in our bodies, have myriad perceptions that make sense of the world, and think every abstract thought passing through our minds. Each of these domains is represented by a golden rectangle, the divine proportion connecting us individually to all that is. When we resonate with beauty in others and in the world, feel love in our hearts, and/or know truth in our minds, we are resonating with whom we truly are.

Philosophers and theologians have long recognized that the transcendental qualities of awareness—truth (the Father), love (the Son), and beauty (the Holy Spirit)—are ontologically one, and are thus convertible. For example, whenever we experience love, we simultaneously experience beauty and truth. When we see beauty, we feel it and know it to be true. Truth is made out of love and shines forth in exquisite beauty.

The circle woven within the triangle symbolizes the unity of awareness. Examine your experience and find out if there is any separation between your thoughts, sensations and perceptions. Is there any distance? Do they take place in separate compartments? They do not. Awareness is seamless. Are the three classes of your experience reducible into one? They are not, but who is the one that experiences these three classes of information? The Phi of God is your viewpoint.

The interlaced triangle and circle surrounding the eye have equal areas, symbolizing the equality of emptiness and form. The three classes of information are separated by a ring, which represents the boundary between your interior and exterior representations.

The Latin inscription above the door, "Sapientia ædificavit sibi domum" means, "Wisdom has built a home for itself." Your home is thrice great in bodies, minds and worlds.

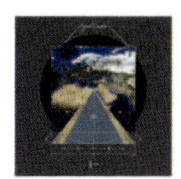

15. Squaring the Circle

When a square and circle are drawn with as close to equal lengths as is practical (8 places after the decimal in my computer), it encodes the divine proportion as shown in two ways, through Φ and $\sqrt{\Phi}$. The wider triangle correlates with the form of the Great Pyramid, and that's why you see stones within and the apex pointing to Giza. By inscribing an equilateral triangle within, the A:B ratio encodes the mathematical constant e, which is of "eminent importance" in mathematics and is widely used in physics, engineering, chemistry, biology, and economics. Of course, e was only rediscovered in the 17th century, but was encoded in the Great Pyramid's geometry nevertheless.

16. Sevens

The Great Pyramid has a base length of 756 feet. Circumscribing the elevation of the Great Pyramid results in a circle with a diameter of 777.7 feet. At the same time, three mutually tangent circles fit within as shown in blue. The two equally sized lower circles have circumferences of 777 feet while the smaller circle's circumference measures 7777 inches! Amazingly, two golden rectangles' primary division at the divine proportion anchors the base of the pyramid within the circumcircle. I inscribed an equilateral triangle within the pyramid and integrated a human form to illustrate how the seven chakras resonate with all the sevens in the design. The edge of the equilateral triangle measures a total of 6666 inches on each edge. Clearly foot and inch repetitive digits resonate with the Great Pyramid's dimensions. How this came to be is a great mystery.

17. Core

The heart is the core of the human energy vehicle, with three chakras above and three chakras below. Our hearts mix cosmic and telluric influences in a crucible with the potential to alchemically transmute our lower fear-based desires and drives into higher radiant manifestations of love. Two divinely proportioned spirals depict this energetic intermixture, which inform and symbolize the human experience.

18. Epiphany

I had an epiphany (e π, Φ, in i) just before publishing this book—simply drawing five circles is enough to encode four fundamental mathematical constants. If drawing by hand, the first step is to open a compass and draw a circle, calling its radius 1. Without changing the spread of the compass, draw a second circle with its center on the circumference of the first circle to create an almond shape between them (also known as the *vesica piscis*). Next draw a vertical line with a straightedge through the two axial points of this almond and a perpendicular line through the two center points of the circles you've just drawn. The right-angled crossing of these lines forms the center of the eye in the illustration.

Place the point of the compass at the intersection of the crossing lines at the center of the eye, adjust the spread to the half-width of the vesica piscis (1/2) and draw a circle. Four red dots show the cardinal points of this circle. Place the compass point on the top red dot and adjust its spread to encompass the distance to the left dot. Draw the upper circle (diameter A). Finally, move the point of the compass to the bottom red dot and draw the symmetric lower circle.

The distance A+B approximates π because it precisely equals $\sqrt{2} + \sqrt{3}$. The degree of inaccuracy (<0.1%) is less than the thickness of the circles. The distance (A+B) x E approximates Euler's number e (to 99.8%) because it equals $\pi\sqrt{3}/2$. The ratio C/D is exactly Φ.

This construction has all the poetic ingredients to comprise the most beautiful theorem in mathematics, $e^{\pi i} + 1 = 0$. Euler's number e and π have already been identified but the imaginary unit i is homonym to "eye." I've illustrated the eye's iris with an inverted view of our Sun, the macrocosmic all-seeing eye. The pupil is a void representing zero and the width of the eyeball is 1. Finally, and perhaps most importantly, the divine proportion weaves the epiphany together.

19. Golden Template

This view of the Sphinx is as close to a front elevation as I have found, affording a geometric comparison between the Sphinx and the Khafre pyramid behind it. The geometry I have superimposed over the photo can be constructed with a compass and straightedge only. Three sets of paired circles form nested divine proportions, as shown by the golden rectangles. The pyramid slope is approximated by the wider triangle surrounding the central equilateral one that was used as a template for the headcloth slope on the Sphinx. The headcloth has distinctive, straight-edged folds that imply an apex point above the head, echoing the apex of the pyramid behind the Sphinx. The center of the vesica piscis pinpoints the third eye of the pharaoh carved on the Sphinx.

I conclude that the same geometric template was used for both the pyramid and the Sphinx. The pyramid resonates with the human form and the divine proportion is the bridge between microcosm and macrocosm.

20. Birthplace of Isis

The simplest way of seeing how a 3:4:5 triangle encodes the divine proportion begins by drawing the triangle in absolute units and then scaling it down by 33.333…%. Next draw the largest circle that will fit along the edge as shown (this turns out to have a diameter of 1). A line drawn from the opposite corner of the triangle through the center of the circle through to its other side has an exact length of Φ.

I was inspired to place a 3D model of the Earth inside the circle and to orient it so that the Earth's axis of rotation ran vertically. I rotated the planet on its axis until Egypt came into view. I drew a horizontal line from the end of the Φ line across the Earth and observed that it marked the latitude of Dendur, which is known as the birthplace of Isis. I see this as particularly appropriate, as Isis was daughter of Earth and Sky and was herself the goddess of Nature, a fitting resonance with our beautiful planet.

21. 3:4:5 Moon Earth

The first Pythagorean triple, the 3:4:5 triangle, encodes the correct Moon:Earth proportion of 3:11 when shown mirrored twice in the top illustration.

The 3:4:5 triangle also encodes the axial tilt of Earth with respect to our orbital path around the Sun, the ecliptic. 66.6° is the actual angle from the Earth's pole, but the complementary angle (90 − 66.6° = 23.4°) is often used to measure the Earth's tilt from the ecliptic to the equator.

22. Body Temple

I updated half of Leonardo da Vinci's famous depiction of the body so that the diagram represents both men and women equally. Corresponding golden rectangles touch each other at the apex of the triangle whose slope approximates the Great Pyramid. Two lines along the primary divisions carry the divine proportion down to where they intersect at the navel, the center circumscribing our physical forms. This diagram suggests that the human body is the true temple on which the Great Pyramid is based.

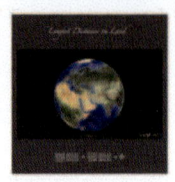

23. Location, Location, Location

Our planet's longest distance on land runs from Western Africa through the Great Pyramid to Southeast Asia. I discovered that the Great Pyramid is located at the Φ point on this longest terrestrial track, given current sea levels. The Great Pyramid appears to have been sited with the divine proportion in conjunction with extremely accurate planetary knowledge. Reality contradicts what is generally known about ancient Egyptian material culture. I believe the Great Pyramid still has much to teach humanity about our past and about the possible existence of non-physical beings (known formerly as gods). The divine proportion is no small part of that lesson.

24. The Nile

The world's longest river symbolizes wisdom, and wisdom always flows from the heart to the head, not the other way around.

The source of the Nile in the illustration corresponds with the heart chakra. The meandering river represents the sinuous flow of energy from the heart to the third eye, which is the precise location of Giza and the Pyramids. The length the average river takes compared to the straight-line distance from its source to mouth is equal to π / Φ. See www.pimeariver.com for more information.

Beautiful numeric symmetries abound when inscribing a unit circle within the Great Pyramid's form. The golden rectangles show that the average height of the human body and the location of the heart within are both divinely and simultaneosly proportioned with respect to the pyramid form.

25. Human Development

Jean Gebser, Clare Graves, Don Beck, Ken Wilber and many other developmental psychologists show that cross-culturally, the growth of human consciousness evolves through a linear sequence of clearly defined stages. Each successive stage adds new world-views, preferences and purposes that transcend and yet still include past perspectives.

I illustrated the developmental progression from the point of view of one's inner experience following a golden spiral. The human boundary of concern naturally expands from one's mother to the ego, outward to family, tribe, and ethnic group, eventually includes the world of international trade and communication, sensitizes to the planet including all of its life, and ultimately embraces the entire cosmos including all previous stages of human development.

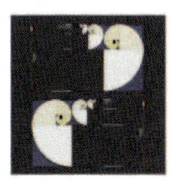

26. Golden Timelines

The divine proportion divides epochal and cultural time into significant milestones. The top progression fits entirely within the bottom progression, namely within the top and final stage labeled "Humans." The golden rectangles give a sense of the exponential nature of time, i.e. colonial photosynthetic bacteria didn't start to leave macrofossils until 1/Φ (~62%) of the time between the formation of the Earth and the present.

The next milestones occurred in an approximate progression of powers of 1/Φ in time, namely multi-cellular life, the emergence of animals, then vertebrates and now us.

Human history since the late glacial maximum and the restart of civilization follows the same pattern starting with the domestication of animals, the invention of the wheel and later writing, to the rise of Greek and Roman city states, to the end of antiquity and beginning of the middle ages, and now to the present moment.

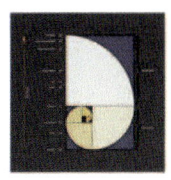

27. Golden Scale

It is estimated that the worldwide average height for adult human males is about 1.72 m, while the worldwide average height for adult human females is about 1.58 m.[1] Therefore, the worldwide average adult human height is approximately Φ meters.

The charge radius of the proton is 8.775×10^{-16} m, which is 16 orders of magnitude smaller than our human scale. The entire universe has a radius of approximately 14 billion parsecs (4.3×10^{26} m)[2], which is 26 orders of magnitude larger than the human scale. The full gamut of scale is thus 16 + 26, or 42 orders of magnitude from a proton to the entire universe. Amazingly, 42 / 26 ≈ Φ, so it logically follows that humans are divinely proportioned in the scale of all things.

28. Spiral Galaxy

This typical spiral galaxy's disc (Messier 51) appears almost perpendicular to Earth's viewing angle. From this perspective, two golden spirals (one positive and one negative) can clearly be seen to inform its macrostructure. All golden spirals are scale-invariant, meaning they have the same form at all scales.

---

[1]. https://en.wikipedia.org/wiki/Human
[2]. http://bit.ly/1Ooyt51

29. Saturn Earth Proportions

Two golden rectangles separated by an equilateral triangle of edge length Φ come together to encode important relationships between Earth and Saturn. The 15-pointed star resonates perfectly with the golden rectangles as shown by the four anchor points marked with red dots, two of them exactly at the Φ points. Saturn's diameter fits on the outer points of the star while the Earth fits within the inner star.

John Martineau pointed out in A Little Book of Coincidence that the 15-pointed star proportions the Earth and Saturn such that it correctly encodes their relative sizes and orbits.

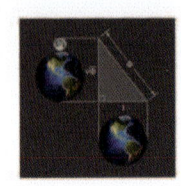

30. Earth Moon Proportions

If the diameter of the Earth is 1, then the combined diameters of Earth and Moon is equal to √Φ (99.9% accurate). The golden ratio is the only number that when added to unity equals itself squared (Φ + 1 = Φ^2). I realized that this naturally suggests a right triangle with sides of √Φ, 1, and Φ by the Pythagorean theorem.

My conclusion is the Earth and Moon are divinely proportioned. Nothing is accidental about this arrangement.

31. Basis of the Metric System

It is well known that the metric system was initially devised by dividing the distance from Earth's equator to the North Pole along the Paris meridian into 10 million parts called meters (which equals 10000 km). We later learned with GPS satellites that this measure is only 99.9% accurate.

I recognized that Earth's polar diameter is 10000 x √Φ km (99.9%) and that the combined diameters of Earth and Moon are equal to 10000 x Φ km (99.9%). Contrary to popular perception, the metric system is not arbitrary but is intrinsically encoded in the Earth-Moon system.

The current definition of the meter is based on the speed of light, which is itself measured with the meter. This is an example of circular reasoning and is literally a logical fallacy. In my opinion the meter should be rooted in the dimensions of the Earth, Moon and the divine proportion, which are already in near perfect resonance.

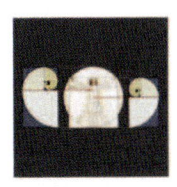

32. Golden Human

Leonardo da Vinci depicted the human body in both a circle and a square, based on the ancient canon of proportion found in the writings of Vitruvius. I perceive the circle, as an infinite number of points equidistant from a center, represents spirit's descent into the human form.

The square's four equal sides represent the corners of a typical building, or any land surveyed using the cardinal directions. The square therefore symbolizes the finite, earthy aspect of the human condition. The circle and the square express the essence of the human vehicle—we are made of equal parts Sky and Earth.

Illustrating the circle and square's divine proportions with golden rectangles reveals their precise alignments with the heart and navel, respectively. The heart is the center of the energetic body and the navel is the center of the physical body. Your subtle energetic body connects you with higher levels while your gross physical body grounds you in the Earth. The ankh as symbol of life, is icon of this sacred marriage. The divine proportion marries Sky and Earth in the human vehicle.

33. Golden DNA

DNA holds the blueprint not only of your body, but of life itself. The molecular structure of deoxyribonucleic acid resonates with the divine proportion in multiple simultaneous ways. The axial view at the top of the illustration shows that the structure's plan is arranged as a decagon, a structure imbued with divine proportions, only two of which are shown given a radius of 1.

In elevation, the golden rectangles show on both sides how the double helix's major groove is structured relative to its minor groove—with the divine proportion. In addition, the overall pitch of the spiral (one revolution is 10 base pairs) compared to its diameter is likewise an expression of Φ (99.9%).

DNA is organized into nucleotide triplets, called codons, which specify which amino acid will be added next during protein synthesis. Geneticist Jean-Claude Perez discovered that the frequency in which codons appear in the human genome are strongly linked to Φ and the musical fifth (3/2). His scientific paper, "Codon Populations in Single-Stranded Whole Human Genome DNA are Fractal and Fine-Tuned by the Golden Ratio 1.618" was published in the journal of Interdisciplinary Sciences: Computational Life Sciences (September 2010, Volume 2, Issue 3, pp 228-240) and is linked here: http://bit.ly/1Fwt9FB. For a layman's description of Perez's paper see http://cosmicfingerprints.com/mathematics-of-dna/

I conclude that the code of life is truly a multidimensional embodiment of the divine proportion.

About the Author

Scott Onstott is a writer, trainer, filmmaker and artist. He has a degree in Architecture from the University of California at Berkeley and has written a dozen technical books on architectural visualization software. He is the creator of the *Secrets in Plain Sight* video series and is author of Quantification and Taking Measure: Explorations in Number, Architecture and Consciousness. Scott grew up in California and now lives in British Columbia with his wife and son.

Visit www.secretsinplainsight.com and www.scottonstott.com for more information.

Printed in Great Britain
by Amazon